Dreaming Ecology

NOMADICS AND INDIGENOUS ECOLOGICAL KNOWLEDGE,
VICTORIA RIVER, NORTHERN AUSTRALIA

Dreaming Ecology

NOMADICS AND INDIGENOUS ECOLOGICAL KNOWLEDGE,
VICTORIA RIVER, NORTHERN AUSTRALIA

DEBORAH BIRD ROSE

EDITED BY DARRELL LEWIS
AND MARGARET JOLLY

Australian
National
University

ANU PRESS

MONOGRAPHS IN
ANTHROPOLOGY SERIES

For all of Debbie's teachers

Australian
National
University

ANU PRESS

Published by ANU Press
The Australian National University
Canberra ACT 2600, Australia
Email: anupress@anu.edu.au

Available to download for free at press.anu.edu.au

ISBN (print): 9781760466275
ISBN (online): 9781760466282

WorldCat (print): 1422761637
WorldCat (online): 1422764418

DOI: 10.22459/DE.2024

Cover design and layout by ANU Press. Cover photograph: Yarralin, November 1980
by Deborah Bird Rose.

This book is published under the aegis of the Anthropology in Pacific and Asian Studies
editorial board of ANU Press.

Publication of this book has been supported by the ANU Vice-Chancellor's Strategic Funds
for Flagship Titles at ANU Press.

Aboriginal and Torres Strait Islander readers are advised that this book contains images
and voices of deceased persons.

Contents

List of Illustrations

Figures

Maps

Tables

Notes for the Reader

Throughout *Dreaming Ecology*, the editors have used Debbie's book *Dingo Makes Us Human* as a guide for the spelling of Aboriginal words and names. We have capitalised Indigenous and Country, in line with common usage in the twenty-first century, and in reference to the relationship Aboriginal people have with their Country. Dreaming and Law are also capitalised throughout. Inconsistencies with spelling, such as the name of Big Mick Kangkinang, which has been spelled variously as Kangkinang and Kankinang, have been standardised—in this case to Kangkinang. The *Macquarie Dictionary* has been used as a source for Australian English.

In relation to referencing, bracketed references have been used for citing published sources. In some cases, Debbie cited quotations through secondary sources, rather than their original publication. These have been left as she originally wrote them. All primary sources (Debbie's teachers) have been included as footnotes and a list of primary sources has been added to the bibliography.

Long indented quotes by Debbie's teachers are primarily in Aboriginal Kriol with editors' notes in square brackets to enhance readers' understanding. Indented quotes from secondary published sources follow convention and use the same font as the rest of the text. We have deferred to the Atlas of Living Australia (www.ala.org.au) as our definitive guide for scientific species names. Italics have been used to indicate words from an Aboriginal language, with the exception of the names of people, spirit beings, places and the collective names of people. While these are not italicised, they are always capitalised. For example, Jimaruk (Dreaming ancestor), Jimaruk billabong (place made by a Dreaming ancestor), but *jimaruk* (water snake).

Following convention, the scientific names for plants are italicised and the first letter of the first word is capitalised. For example, *Melaleuca argentea,* but *Vigna lanceolata* var. *lanceolata* (note 'var.' is not italicised).

Scientific names used generically, such as brachychitons or acacia, are not italicised or capitalised. The Aboriginal names for plants are all italicised, although they are not capitalised, for example, *pakali*.

Many of the images in *Dreaming Ecology* were taken by Darrell Lewis and are used with his permission and the permission of the subject/s. Appendix 1 contains a letter from Brian Pedwell, Mayor of Walangeri Ward of the Victoria Daly Regional Council, in which he explains the process involved in getting permission to use the photos that contain recognisable images of individuals. Permission has been gained to use images other than those provided by Darrell Lewis.

Acknowledgements

Publishing *Dreaming Ecology* several years after the death of Debbie Bird Rose has been a process of dedication and love. We thank Chantal Jackson for keeping faith with her mother's wishes that this fine book be published to complete an envisaged trilogy. The editors Darrell Lewis and Margaret Jolly want to thank each other for sustaining a happy and harmonious working relationship in the complex process of developing this work for publication. We are grateful to Karina Pelling of CartoGIS at The Australian National University (ANU) for finessing regional maps and bio-eco maps of remembered walkabouts done by Debbie's teachers long before she came to Yarralin, to such a high professional standard. We are also extremely grateful to Carolyn Brewer for her excellent work in editing, standardising the text and chasing up some elusive bibliographical details. We warmly thank Beth Battrick for her prompt and meticulous work on final copyediting, styling for ANU Press and creating the index. We thank the National Library of Australia for permission to publish a copy of Thomas Baines's early engraving of a boab tree. We thank Matt Tomlinson and the Anthropology Board of ANU Press for the invitation to submit this as a Flagship title and to the ANU Vice-Chancellor Brian Schmidt for funding Flagship titles of ANU Press. Margaret Jolly also thanks the School of Culture, History and Language in the College of Asia and the Pacific at ANU for research support over decades and into her retirement.

We are grateful to both Richard Davis and Thom van Dooren for close and insightful readings, for excellent suggestions to which we have responded and for revealing themselves to us. We are especially pleased to be able to include a suite of photographs of both people and Country (mostly taken by Darrell Lewis) as Debbie had planned and also to link this volume to a gallery of other photographs—most in splendid colour. We thank the current mayor of Yarralin, Brian Pedwell, for confirming permissions given at the time of research to publish photographs of people, many of whom have now passed into Country (see Appendix 1). But our deepest and most humble thanks must be to Debbie's teachers who were the source of the profound knowledge and moving stories offered in this book. We see this as a return gift to present and future Indigenous generations and their gift to a broader public audience in what is a critical moment for Indigenous concerns in Australia.

Acknowledgements by Deborah Bird Rose

The majority of the research was carried out in the communities of Yarralin and Lingara in the years 1986–1991, with less detailed research with people in the communities of Daguragu, Kalkarindgi, Pigeon Hole, Gilwi, and Timber Creek. These communities are all located in the Victoria River Valley of the Northern Territory of Australia. The methodologies will be discussed in the following chapters, as will many of the social boundaries and social groups through which people in this region identify themselves and their connectivities.

Much of the research presented here was funded by the Australian Institute of Aboriginal and Torres Strait Islander Studies, and I remain grateful for their committed support for my research. The final portions of the research, and all of the writing, was accomplished in my position at the Centre for Resource and Environmental Studies of The Australian National University. I am grateful to several botanists whose impact is felt throughout this book. Thanks to Neville Scarlet for introducing me to the joys of ethnobotany, to David Cooper for assisting me with my first ethnobotanical work in the Victoria River District, and to Jock Morse for assistance in analysing the information back in Canberra.

My debts to my Aboriginal teachers are evident throughout the book, but it is always worth saying again: their hospitality, generosity, teaching, and patience have taken my understanding, and indeed my life, in directions for which I will always be grateful.

My loving thanks, as ever, go to Darrell Lewis who always gets up to light the fire.

Prelude 1.
Dreaming Ecology: Bringing to Fruition

Chantal Jackson

I knew that Mum had wanted to complete this book, that it had been underway and on the back burner for many years. When she became ill with cancer, her first priority was to complete *Shimmer*, which she miraculously managed just a few weeks before she died. Her hope had been to have time to complete the other books she had a longing and a commitment to finish. This was not possible.

A few months before she died, I asked Mum to walk me through her 'stuff' and let me know what she wanted done with it; research materials mostly. She showed me a ring folder that had this manuscript in it and she said, 'It's nearly done, it mightn't take much work to get it to publication; the chapters are all there.' I shelved this in the back of my mind for nearly a year. But when I brought her research materials back to Canberra and began sorting them for the Northern Territory Archives, I came across the folder and realised I couldn't deal with it then, so put it aside.

Some months later, it was still tapping me on the shoulder. I knew that I did not have the time or the expertise to finish the manuscript for Mum, and so I reached out to her colleague Dr Peter Read, who suggested I talk to Dr Tom Griffiths. On Tom's suggestion, I contacted Dr Margaret Jolly, a colleague from Mum's Australian National University days who kindly stepped in and offered to help. And when I spoke to Dad, Dr Darrell Lewis, he also stepped up and stepped in. I am so grateful to these two people for bringing this work to fruition. It has been a huge amount of work to edit the manuscript, compile a bibliography (which was missing), locate and in some cases create maps, source diagrams and work out how best to bring this work into the world.

I see *Dreaming Ecology's* importance as twofold. First, it is crucial that the commitment Mum made to her teachers at Yarralin and other communities, as part of their generous sharing and their desires for what would happen with what they taught, should be brought into the world as best as we now can. As I recall Mum talking about her role, Yarralin people saw her as a conduit, a bridge between blackfella way and whitefella way. She was 'read and write' (literate), she could listen and understand at least some things, and she was trusted to transmute what was taught into whitefella language. She could give 'Aboriginal way' a place in the lexicon of *kartiya* (whitefella) ways of sharing knowledge. As I understand it, Yarralin people had a longing to be known and understood and to share.

Second, I joined Mum and Darrell—as he was to me then—at Yarralin in 1981–82 for eight months of her two-year research and accompanied them on many subsequent trips. I have ongoing connections to the community. I have spoken of *Dreaming Ecology* with Aileen Daly and Cerese Young, daughter and granddaughter of Daly, one of the men Mum worked with, and they are excited and eager to see the book and to have it published. And while they and many others may not yet have the 'read and write' skills that much of this book requires, there is the fact of its existence in the wider world and the photos which will be pored over and bring memories and connection. It is also here for future generations, always coming, who may want this link to the people who are gone and to a landscape—to Country—now deeply altered by white invasion and settlement.

With the work Margaret and Darrell have put into bringing this book to publishable quality and ANU Press's wonderful online publication, this book is our best attempt to fulfil her promise to her teachers and to give back to community.

Chantal Jackson
Literary executor for Deborah's Indigenous work
February 2023

Prelude 2.
Debbie and Yarralin: The Early Years

Darrell Lewis

Debbie came from the USA to Australia in July 1980 to carry out PhD research into, as she put it, 'Aboriginal identity'. At the time, her plan was to find a community with a strong traditional culture, remain there for a year and then return to the USA. After a week or so with friends in Sydney she arrived at the Australian Institute of Aboriginal Studies (now the Australian Institute of Aboriginal and Torres Strait Islander Studies, or AIATSIS) in Canberra, which was partly funding her research. This is where we first met.

Debbie's first attempt to find a field location was to contact the Mowanjum community, near Derby in Western Australia, but her request to carry out research there was denied. This was, of course, a great disappointment, so I sat with her and went through what I knew, from firsthand experience or hearsay, about Aboriginal communities in the Northern Territory. Some were located in very harsh desert environments. Others had long been influenced by missions, which tended to weaken or modify traditional culture. Yet others were close to a town or roadhouse and consequently endured severe alcohol problems. Many had a combination of these influences and problems.

My recommendation to Debbie was Yarralin, a community on Victoria River Downs Station where I had worked on sacred sites' documentation in 1975 and on land claims in 1977. The location was a long way from the nearest town and had never had mission activity. It had wonderful scenery, rich biodiversity and large natural waters, and the people were friendly and culturally strong. A friend, Ros Fraser, told her then partner, Jack Doolan, about Debbie and her search for a field location, and mentioned Yarralin as a possibility. Jack was the Member for Victoria River in the Northern Territory Legislative Assembly and had been actively involved in the establishment of Yarralin community in the 1970s. On Debbie's behalf Jack spoke with Yarralin community leaders, and their response was positive.

Debbie arrived at Yarralin in September 1980 where she was warmly welcomed and given a house to live in. This 'house', the same as was lived in by most community members, consisted of a concrete-floored, corrugated iron-clad shed, with a verandah, a single power point and a tap in the yard. Showers and other facilities were communal and all cooking was done on an open fire. Initially, a constant stream of Yarralin men came to Debbie and began to teach her their culture—mythology, ecological knowledge, kinship system and language—and their history and life stories. Many of them politely suggested that, if she wanted, she could come and stay in their camp. Debbie equally politely declined their invitation, telling them she already had a husband.

Interestingly, nearly a year passed before the Yarralin women began seriously to interact with Debbie. They told her that they thought she would find the climate and living conditions too hard and wouldn't stay long, so there was little point in establishing relationships with her. Eventually, Debbie formed strong and ongoing relationships with many of the women.

Some of Debbie's teachers, both men and women, had grown up in the bush. Others grew up in station 'blacks' camps'. All had been on walkabout in the wet seasons, foraging for food and learning about Dreaming sites. All had been observers and participants in ceremonial life. In the case of men, most had been stockmen, mustering cattle all over the stations and in the process learning the location and significance of Dreaming places from older men. In other words, all of Debbie's teachers were strong in the 'Law', possessed of deep knowledge of ceremony, mythology, the location and significance of Dreaming sites, and all other aspects of traditional culture.

I joined Debbie in Yarralin in November 1980. My self-appointed role was to provide backup—to collect firewood, to drive and navigate on forays into the bush or to other communities, to photograph community life and to help Debbie gain an understanding of the local creole, geography, settler history and wildlife. This enabled her to fully concentrate on learning from Yarralin people and to fully immerse herself in community life, including participation in ritual events. It didn't take long for Debbie to realise that her original plan to stay for 12 months was grossly inadequate; we were in Yarralin for nearly two years and later spent other prolonged periods there, working on a number of regional land claims, helping register sacred sites with the Northern Territory Sacred Sites Authority (now the Aboriginal Areas Protection Authority, or AAPA), and on other research projects.

From Debbie's initial research and later projects came her prize-winning books, *Dingo Makes Us Human* and *Hidden Histories,* the first two of a planned trilogy. The third is this volume, *Dreaming Ecology.* The book reveals the extraordinarily dense and complex interconnections, rights and responsibilities that ordered life in the traditional Aboriginal world, and the philosophical and cosmological underpinnings of these relationships—between people and the natural world, between one Aboriginal group and another, and relationships between different species and/or natural phenomena.

To Australia's great benefit Debbie's initial plan to return to America after 12 months' field research never eventuated. She instead became an Australian citizen and spent the rest of her life here. Her passing is a major loss to this country, but her numerous writings dedicated to social and ecological justice, and to helping bridge the gap between Aboriginal and settler Australians, are an ongoing gift to this country, and to the wider world.

Dr Darrell Lewis
Adjunct Senior Lecturer
Department of Archaeology and Palaeoanthropology
University of New England
Armidale
February 2023

Prelude 3.
Deborah Bird Rose —
Ahla Tyaemaen

Linda Payi Ford

Introduction

It is with honour that I have known the great works produced by Ahla Deborah Tyaemaen Rose. Ahla Deborah Tyaemaen Rose is always there for me, and she will always be with me in spirit. This is my story of how Ahla Deborah Tyaemaen Rose shared her sweat on my Country, *Kurrindju. Putj murikimiya woewoe, Ma!*

Historical background

The reality of the history of my land rights and sovereignty is complicated and painful. For countless generations, Rak Mak Mak Marranunggu, DjataDjat, Rak Numala and Marrawulgat people, as described by my Kagal or Uncle Fred Waters while giving evidence in the Finniss River Land Claim (1980–81), belonged to our traditional lands, fresh and salt water, and the cosmos. Today I continue to acknowledge my ancestors accordingly and will refer to us or myself as Mak Mak in this paper.

The first permanent white settlement in our proximity was established by George Goyder and his surveying team at Port Darwin on 5 February 1869. In the early 1900s an Aboriginal Reserve was established on the Finniss River.

The Mak Mak—my old people—incurred many injustices. First, their land was stolen by the colonisers and some of our ancestors murdered. Later, some of their children were forcibly removed, along with some of the adults as well.

In more recent times my senior Elders—George Wigma, Fred Waters, Leo Djekaboi, Peter Melyen, Bilawuk Kirol, Pam or Pandela Clayton and Nancy Daiyi—had to stand up against coercive actions by the Department of Native Affairs, such as the removal of their children who are members of the Stolen Generations. From the late 1970s onwards, they also had to contend with the Northern Land Council. My senior Elders' understanding of the Westminster system and of both written and spoken English was extremely limited. Most of them could barely read or write their own names. *Keh!* This is why I share my story of our struggle for social justice.

It was not until the late twentieth century that a move to restore some of the Indigenous land rights was attempted. In 1973, in response to an inquiry by Justice Woodward, a statutory body known as the Northern Land Council was established in the Northern Territory. The Northern Land Council's role was to assist in claims to land by ascertaining the views of Aboriginal peoples and advocating for our interests. During the 1976–81 Finniss River Land Claim process, several groups had disputed ownership of the Delissaville, Wagait, Larrakia Aboriginal Trust, and nearby Crown land. The Rak Mak Mak Marranunggu people were recognised as the Traditional Aboriginal Owners under the *Aboriginal Land Rights (Northern Territory) Act 1976* of Areas One in 1991 and Two in 1993 of the Finniss River Land Claim. These sections of the claim are now known as Gurrudju Aboriginal Land Trust.

At the time, my mother, Ahla Nancy Daiyi, contested the Northern Land Council's decision to recognise other clan groups as Traditional Aboriginal Owners. The Federal Court heard the case in 1991 with Justice Olney, the Aboriginal Land Commissioner, presiding. He overturned the Northern Land Council's 1981 decision. Justice Olney found that the Northern Land Council had not appropriately examined all the evidence provided to them by our old people, the senior Elders of the Mak Mak. He found that the Council had made the wrong decision in 1981 in the Finniss River Land Claim. Justice Olney ordered the Northern Land Council's decision to be set aside and directed the Northern Land Council to conduct new hearings in what became known as the Wagait Dispute Hearings 1993–94.

Meeting Debbie Rose

Beginning with the Finniss River Land Claim in 1973, Rak Mak Mak Marranunggu worked with many anthropologists over the next four decades. In 1993 we invited an anthropologist and a lawyer to attend a meeting in Batchelor to discuss with them their interest in working with us on our land claim for the Delissaville, Wagait, Larrakia Aboriginal Land Trust, which the Mak Mak referred to as The Reserve or the Big Wagait and the Little Wagait. The Little Wagait is a section of Aboriginal Land Trust that does not join with the Big Wagait but is located east of the Big Wagait. Deborah Bird Rose came to Batchelor and announced at our initial meeting that she preferred to be called Debbie. She had come to meet with us and to see who we were. Later that year, Debbie notified the Northern Land Council that she would act as the Rak Mak Mak Marranunggu anthropologist in the forthcoming land claim hearing.

We were very excited to accept Debbie as our anthropologist and she was invited to meet with the Mak Mak family at our White Eagle office in Batchelor to plan our case to present at the hearing for the Wagait Dispute. Debbie spoke to all of us and wanted to assess what we thought the land claim meant to us and to prepare for site visits on Country and giving evidence at the Wagait Dispute Hearing. She cautioned it was going to be lengthy and intense—that, it certainly was!

The Wagait Dispute Hearings

The hearings were to consider the arguments of five competing clans which were disputing ownership. They were chaired by a senior local anthropologist and overseen by a Queen's Counsel from the Justice Department, with a panel of members representing the Full Council of the Northern Land Council. The committee's determination was announced in 1995 when the Mak Mak people were declared Traditional Aboriginal Owners of the major part of the disputed lands—the E2 of the Delissaville, Wagait and Larrakia Aboriginal Land Trusts.

Bestowing a Mak Mak name

In the course of preparing for and engaging in the Wagait Dispute Hearings, we all developed a good relationship and connection with Debbie. My Ahla (mother) Nancy Daiyi, a senior Elder of Mak Mak, bestowed Debbie with her deceased sister's name, Tyaemaen. This meant that Nancy had accepted Debbie as her classificatory sister. Therefore, I was now to refer to Debbie as Ahla Tyaemaen and my Mak Mak kinship roles and responsibilities with Ahla Tyaemaen meant an extension to our relationality.

As time went by, my relationship with Debbie grew and formed a solid bond between us. This was an essential quality embedded in our 'daughter–mother' relationship. It was special as it was authorised by my Ahla Daiyi through a cultural iterative process of drawing interconnections between discrete categories of knowledge, knowing, being and doing. It allowed my relationship with Ahla Debbie Tyaemaen Rose to grow respectfully, professionally and personally. I viewed Ahla Tyaemaen through a Mak Mak lens as the first anthropologist who addressed our social justice issues using the Northern Territory Aboriginal Land Rights legislation.

Full circle

In 1991 my husband Mark and I returned to my hometown and birthplace at Batchelor in the Northern Territory. I commenced my professional tertiary teaching at Batchelor College as an academic lecturer. In 1993, I completed my Graduate Diploma in Special Education, awarded by the Warrnambool College of Advanced Education (later amalgamated with Deakin University). Professor Pat Varley travelled to Batchelor to deliver my Warrnambool College of Education testamur at the Batchelor College's graduation ceremony. My Wangga and Wali ceremony family from Wagait and Daly River danced me up to receive my degree.

Prior to my graduation ceremony I had never imagined my life as an Aboriginal scholar. At the time, I was teaching in the tertiary education programs at Batchelor College (now called the Batchelor Institute of Indigenous Tertiary Education), applying the 'Both Ways Philosophy' developed by Yolngu from East Arnhem Land, which was being appropriately advocated by colleagues at this time. It was an exciting time and place to be working in tertiary education.

I recall having numerous discussions with Debbie about discourse and pedagogy, about Aboriginal knowledge, languages and cultural practices, and what this meant for Aboriginal people. This was especially informative since I was then working as an academic at Batchelor College and past practices were being challenged. Those discussions opened my mind's eye to endless possibilities for Aboriginal people and for myself. In 1995 my father Maurice (aka 'Max' Sargent) and I were having a yarn about Debbie, and he mentioned how Debbie had taken an interest in my career and had taken me under her wing. I agreed and confirmed that I would be respectful of our relationship and that I would approach Debbie to discuss career options with her. At the time Dad was very unwell and he passed on 24 June 1995, alas, just before the Wagait Dispute Hearing Committee handed down its decision.

I was keen to enrol in my Master of Education in 1995 at Deakin University. Ahla Tyaemaen encouraged me to apply for the inaugural Stanner Scholarship at The Australian National University. My application was successful, but I ceased the scholarship in early 1996 because I had become pregnant with Chloe Ngelebe Ford. Three years later, I graduated with my Master of Education from Deakin University. This was the year Emily Tyaemaen Ford, my second daughter, was born. My husband, Mark, and I believed we were truly blessed. We had our own family and two beautiful daughters.

In 1998, I enrolled in a doctoral degree part-time at Northern Territory University but, in 2001, I transferred my doctoral studies to Deakin University. I had resigned from Batchelor and accepted an academic lecturing position at the Northern Territory University in the Faculty of Science, Information Technology and Education. Northern Territory University became Charles Darwin University in 2003.

Ahla Tyaemaen continued to stay in close contact with Ahla Daiyi after the Wagait Dispute Hearing. Throughout this period Ahla Tyaemaen had been drafting the book *Country of the Heart* with us and an American photographer. She organised slide nights for editing purposes for the book at our place in Batchelor or at her place in Tiwi in Darwin's northern suburbs. The first edition of *Country of the Heart—An Indigenous Australian Homeland* was published in 2002 by the Aboriginal Studies Press at the Australian Institute of Aboriginal and Torres Strait Islander Studies in Canberra. All the co-authors celebrated the amazing event with the first-named author, Deborah Bird Rose.

The power of publishing

In the late 1990s, Ahla Tyaemaen relocated to Canberra, so we didn't see each other as often as we would have liked. Still, she invited me and family members to do guest lectures on a regular basis at The Australian National University. She also organised an installation of the Mak Mak Marranunggu in the *Tangled Destinies* exhibition at the National Museum of Australia in 2002. This had a huge impact on my views on what museums and archives were and how they housed historical records and influenced Australian and overseas visitors who would check out their collections.

Ahla Tyaemaen had indeed opened my eyes to scholarly work. I felt there was honour and respect in her approach to working with Australia's First Peoples. This is evident in this book, *Dreaming Ecology*, in how she celebrates the knowledge of her Indigenous teachers and sees her words as a bridge for them. This was also illuminated when she shared in the co-authorship for the publication of *Country of the Heart* in 2002 (and its republication in 2011). The unprecedented reciprocation of western 'treasures' in this way with us was nothing short of brilliant. It was from that moment that my view as a higher education scholar about authorising the application of my Aboriginal knowledge in my own research projects shifted. This was mainly in how research and publishing Indigenous knowledge could be achieved. I was no longer the subject but the owner of my published work.

Ahla Tyaemaen had ignited the metaphorical fire within me to address Indigenous knowledge, intellectual property and shared ownership through my teaching and research in higher education, as well as in community development projects. Publishing research changed the way I thought about publishing our stories—as it did for Ahla Daiyi. She would often say, 'If only I could read and write I would write my own book!' She said this often when outsiders visited our Mak Mak Country, and our communities couldn't understand what she was expressing. Later Ahla Daiyi, in collaboration with Ahla Tyaemaen, did publish that book, *Country of the Heart—An Indigenous Australian Homeland*. She was on the front cover, proudly walking on her Country to her favourite fishing spot at Ditjini.

Standpoint in womanhood

The cultural and language barriers between Indigenous and non-Indigenous people were and still are present. This would frustrate Ahla Daiyi and me no end. The relationship that Ahla Tyaemaen had with her sister, Ahla Daiyi, was unique. They could understand each other's standpoint and womanhood in iterative ways. Both women were able to analyse their intersubjective discourses where their standpoint positioning was in their own authoritative domains and rooted in their own individual personal knowledges and perspectives, as well as the power that such authority exerts. Thus, the shared space of understanding their subjectivities and epistemologies, based on their lived realities, generated a nuanced theoretical approach. This was clear in how they navigated the difficulties of their female experiences within spaces which challenged their epistemologies and their feminine discourses over their lifetimes.

They built on their positionalities in the land claim process, a form of knowing through a method of inquiry, described by Martin Nakata as making more intelligible the corpus of objectified knowledge about us as it emerges and organises understanding of our lived realities (2007, 213–16). This was a unique experience for me to observe as it unfolded. I now reflect upon those histories, applying a new lens from my own feminine discourse and genre.

Standpoint theory has allowed me, based on my own lived experiences, to place important values on those experiences where two senior women dialogue as a source of knowledge. Moreover, Brenda Allen (1996) describes how women of marginalised and/or oppressed individuals can support each other. For me, the Wagait land claim was organised as a quest to create more objective accounts of the world.

In this social and political context, both my Ahlas gave voice. Patrice Buzzanell (2003) describes how standpoint theory gives voice to marginalised groups, allowing them to challenge the status quo as outsiders within the status quo, which represents the dominant position of privilege. In the case of the land claim, the dominant position of privilege was held by the Northern Land Council.

As described by Victoria DeFrancisco (2007), the predominant culture in which all groups exist is not experienced in the same way by all persons or groups. The views of those who belong to groups with more social power are validated more than those in marginalised groups. Those in marginalised groups must learn to be bicultural, or to 'pass', in the dominant culture to survive, even though that perspective is not their own.

In 1991, I realised how much Ahla Daiyi really wanted to know more about writing, reading and multiple literacies. She was hungry for more knowledge and to share her knowledge. Ahla Daiyi enrolled in a vocational training literacy course in 1992–93. Ahla Daiyi, Mark and I, and later Chloe and Emily, travelled as a family unit across the world to places where Ahla Debbie Tyaemaen Rose had organised book launches. We went to the Sydney Writers' Festival, the University of Melbourne, Scotland, America and The Australian National University. Then, for the second edition of *Country of the Heart* in 2011, we went to the Brisbane Writers' Festival. Ahla Tyaemaen certainly looked after us in this regard.

Outcomes of high knowledge

In October 2006, I graduated with my doctorate 'Narratives and landscapes: Their capacity to serve Indigenous knowledge interests'. My Ahla Daiyi and Ahla Debbie were very proud. In turn, I was proud of both my Ahlas for their contribution to my rites of passage in both cultures. At the time Ahla Daiyi was unwell. She passed on 4 April 2007. I, as well as loved ones and others, were dealt such a huge blow. Ahla Tyaemaen attended the Ceremonies on Country on the Wagait after Ahla Daiyi's passing. Ahla Debbie continued to work on projects that she was passionate about. She loved the environment, plants and animals. Her words describing Country were profound. Our book, *Country of the Heart*, captured the moods of my Country's spirit.

Debbie's anthropological studies were amazing. In this book, *Dreaming Ecology*, you can feel the heart, intellect and deep respect she brought to her studies of the First Nations Peoples of the Victoria River District. In her later years in academia, her focus was on the anthropology of the environment and the changing climate's impact on the land, fresh and salt water, plants, animals and cosmos. Ahla Tyaemaen supported my Australian Research Council grants and other research projects. I felt blessed to have someone of her standing in academia to be working alongside me, providing a collegial way of thinking, doing and being, while collaborating and advising me on sensitive issues as they arose.

When Ahla Tyaemaen informed me that she was unwell, my family and I were devastated. We often visited her when we were travelling to Sydney University where I was working as an Adjunct Research Fellow at The Sydney Music Conservatorium. My research had taken me, my family, other Mak Mak and colleagues to museums, archives, and private and public institutions across the nation and globally. There are amazing amounts of information stored in archival collections waiting to be read and critically interpreted.

Ahla Tyaemaen passed away in December 2018. What a great loss to the Australian academy and to me. Both my Ahlas' spirits continue to influence my research and are powerful drivers in what I do today. These two giant mothers whose shoulders I stood on were my key family members. They supported me and gave me the strength and confidence to move forward with my research and celebrate in our successes altogether.

Both Ahlas, Daiyi and Tyaemaen, always gave me 100 per cent support and encouraged me to continue to further my education, extend my understanding and to share my knowledge. Ahla Daiyi would always remind me by saying, *'If you don't do it, no one else will!'* The lessons learned from these wise women have nourished my research spiritually and their presence continues to grow through my research projects. Both Ahlas have taught me about the benefits of life on Country, lived realities and connections. This has also taught me that my oral histories, along with historically documented files in museums and archive collections, are there to access as tools to integrate traditional knowledge systems with other knowledge systems. I have developed my own repository with the Pacific and Regional Archive for Digital Sources in Endangered Cultures (PARADISEC). This was a collaborative effort with my colleagues, and my Mak Mak family members.

This story is deep. The knowledge from my Ahlas will continue to be transferred by me, to you. Honouring my Ahlas' knowledge about history provides an account of the past to right the wrongs and to include an Aboriginal perspective. This practice is formalising an alternative lived historical narrative to forge a better future for the next generations of Australians.

The future

The history of our engagement with the Northern Territory Aboriginal Land Rights Act created a milestone, a lived moment in my family's shared histories. Those days will always remain etched in my memory and will always be a memorable period of my then young life. My story potentially offers a pathway to understand the struggle of Aboriginal people of Australia and those of us under the Northern Territory Aboriginal Land Rights legislation.

It is up to individuals to come to terms with their identity and what they wish to do and how they determine their own pathways to aspire to their full potential. The destiny and life of an Aboriginal person is influenced by colonising history, by pain and by grief at the loss we have endured. It is up to us to have a voice and decide how this may or may not contribute to a future with justice to honour the First Nations Peoples' ancestors and Country.

Ahla Tyaemaen's professionalism on behalf of the Mak Mak people enabled us to stand up and have our voices heard. Today we are recognised as Traditional Aboriginal Owners of the lands covered by the Delissaville, Wagait and Larrakia Aboriginal Land Trusts and the Little Wagait Aboriginal Land Trust. This is a small portion of our traditional lands and sea in the Finniss River and Reynolds River region that we fought so hard to claim. It was a significant hurdle to pass over and a milestone never achieved before. It allowed the Mak Mak to be proud of our First Nations' identity.

Sadly, there are many, many First Nations Peoples that continue to struggle with their identity and Country. This is through no fault of their own, but because of colonial practices and legislation of the past that led to the taking of so much that belonged to us on our own lands, sea and cosmos. It cut off and deprived us all of rich connections and histories. The depth and breadth of these rich connections and the losses wrought by colonialism are beautifully described for the First Nations Peoples of Yarralin and Lingara by Debbie in *Dreaming Ecology*.

Linda Payi Ford
Associate Professor
Charles Darwin University
Darwin
May 2023

References

Aboriginal Land Rights (Northern Territory) Act 1976. Online: www.legislation.gov. au/Details/C2016C00111, accessed 5 May 2023.

Allen, Brenda J. 1996. 'Feminist Standpoint Theory: A Black Woman's (Re)view of Organizational Socialization'. *Communication Studies* 47 (4): 257–71. doi.org/ 10.1080/10510979609368482.

Buzzanell, Patrice M. 2003. 'A Feminist Standpoint Analysis of Maternity and Maternity Leave for Women with Disabilities'. *Women and Language* 26 (2): 53–65.

DeFrancisco, Victoria P. 2007. *Communicating Gender Diversity: A Critical Approach.* Thousand Oaks, Indiana: Sage Publications. doi.org/10.4135/9781483329284.

Nakata, Martin. 2007. *Disciplining the Savages: Savaging the Disciplines.* Canberra: Aboriginal Studies Press.

Prelude 4.
Dreaming Ecology:
Reflections

Margaret Jolly

Process

In the months before she died in December 2018, Debbie Bird Rose expressed her desire to her daughter Chantal Jackson that *Dreaming Ecology* might be published. This would complete her envisaged trilogy, based on her long-term experiences with the Yarralin mob, to follow her earlier widely celebrated works, *Hidden Histories* (1991) and *Dingo Makes Us Human* (1992). Based on years of research commencing in 1980–82 with many return visits, the manuscript had substantially been completed in 2003. But she knew that there were still parts that she wanted to elaborate, a full set of references to be finished and maps, figures and photos to be located, inserted and finessed. During those last years and months, Debbie was preoccupied with other work, her book *Shimmer* (2022) and several other studies in environmental humanities and extinction studies as her own strong life force started to ebb.[1]

In July 2021, in the midst of a cold pandemic winter in Canberra, I received an email from Chantal asking me if I might be involved in helping to shepherd the manuscript to publication. I was honoured to be asked since I had long been an admirer of Debbie's work—as an anthropologist who

---

1   For cognate arguments see Rose (1991, 1992, 1996, 1997, 1998, 2000a, 2000b, 2004, 2022). The most up-to-date curriculum vitae for Debbie that we have found is at the following link: unsw.academia. edu/DeborahRose/CurriculumVitae, accessed 5 March 2023. Also see the Wikipedia entry, 'Deborah Bird Rose', Wikipedia, last edited 6 May 2023, en.wikipedia.org/wiki/Deborah_Bird_Rose, accessed 17 August 2023. See also the fine obituaries by Stephen Muecke (2020) and Eben Kirksey (2021).

had been deeply engaged with the Indigenous Australians she had lived and worked with, who practised not a form of evanescent ethnography which extracted data from 'informants' but a person who listened deeply and respectfully to her 'teachers', who celebrated the depth of Indigenous Australian philosophy and who nourished lifetime connections in struggles against the ravages of colonialism to regain Country which had sustained an Indigenous life for 65,000 years.

When I first read *Dreaming Ecology* I found its breadth and depth breathtaking. But I explained to Chantal that, although as a feminist scholar researching gender and the climate crisis in the Pacific, I might have the scholarly background and passion to help bring this important work to fruition, I could not conceivably do this alone. I had not walked on Country in the Victoria River District, I did not know its people, I was not fluent in Aboriginal languages and had not been engaged with the scholarly debates swirling around Indigenous Australian ecological knowledge and practice. So, we would need to find a willing co-editor.

We found that in Darrell Lewis—Chantal's father, Debbie's long-term partner for 27 years during which they lived and worked with the mobs at Yarralin, Lingara and Pigeon Hole. Darrell introduced Debbie to this Country and is a published author of many books on the rock art, environmental changes and the history of pastoralism of Northern Australia.[2] Despite the emotional challenges of their separation later in life and Debbie's subsequent death, Darrell wholeheartedly agreed to be involved. Most of the rigorous textual work of verifying references, adding explanatory interpellations to long quotes in North Australian Kriol and Aboriginal Pastoral English and sourcing the bio-eco maps and figures from Debbie's archives in the Northern Territory, has been his. He collaborated with Karina Pelling of CartoGIS at The Australian National University to fine-tune the maps of the region and the bio-eco maps Debbie created with her teachers. I assisted with some elusive references, resolved complex issues of overlap and duplication arising from successive drafts, added some notes and changed some lingering Americanisms to Australian style. For the most part we have tried to amplify Debbie's original voice, resisting updates and corrections and giving you as reader a sense of how she might have finished and finessed this book if she had been given more time.

---

2   Some of his major works are listed in the references: Lewis (1988, 1996, 2002, 2007, 2012) and with Rose (1988) and with Schulz (1995).

From our first editorial meetings, Darrell showed me a gallery of exquisite photos he had taken of Indigenous people and of Country which complement the text in graphic ways. Debbie had planned for many of these photos to be an integral part of her book's design. By publishing both digitally and in hard copy with ANU Press we are able to include a wide range of these photographs, most of which are in full colour, and to add a gallery of these in a linked website. The fact that all of this is online and free to download realises our strong desire, and that of Debbie, to give this book back to the Yarralin mob. Almost all of Debbie's most important teachers have now passed back into Country. As I write, only Allan Young is still living. But, while deeply mindful of the Indigenous protocols about names and images of those who have passed, we see *Dreaming Ecology* not just as a reciprocal gift from Debbie, but from one generation to another. A return in every sense.

Ancient practices, prescient insights

> It explores a holistic understanding of the interconnections of people, country, kinship, creation and the living world within a context of mobility. Implicitly it asks how people lived so sustainably for so long.

This is how, in 2003, Debbie distilled the significance of *Dreaming Ecology* in a prospectus for publishers and readers. The book celebrates the depth of Indigenous understanding both in the sense of philosophical profundity, but also in terms of the temporal depth of Indigenous knowledge and practice, now known to extend over 65,000 years. But it also offers insights for white settlers, those of us who first occupied Australia from 1788, insights which are prescient given the escalation of the climate crisis, the conversation about the Voice to Parliament in Australia and for the scholarly field of environmental humanities which Debbie helped to create and sustain.

It is hard to distil the core insights of a book bursting with innovative ethnographic and theoretical ideas, but in these reflections, I focus on five themes that Debbie develops in the course of *Dreaming Ecology*: footwalk epistemology, totemism, 'wild' Country and 'double death', absence/ presence, and creation and return.

Footwalk epistemology

As in much of her writing, Debbie offers a grounded and embodied sense of place.[3] Especially in the opening passages reflecting on her long walks across Country with Jessie Wirrpa we can feel the red dirt on her soles, light moving in diurnal and seasonal rhythms, the searing heat of the dry season and swirling dust storms, the torrential flooding rains of the wet, rivers filling and drying out, the changing sensation of winds. Country is alive with the sounds and traces of other creatures—insects, birds, snakes and lizards, kangaroos, turtles, fish. Their presence and calls speak to each other and to humans—telling when plants come into flower and seed, when fish are moving, and where there is an abundance of bush tucker and bush medicine. Humans who pay attention in a full-bodied, multi-sensorial way can hear the messages of Country, alive with far more than a human presence and sentience (Chapter 2, 63–64). Indigenous Australians can now move across Country in trucks and cars, but they celebrate 'footwalking': the feeling of feet moving over the ground, the movement of breath and wind, climbing across hot sandstone, pausing at billabongs and springs, sensing characteristic smells, the prickly brush of spinifex, the bites of ants and mosquitoes and all the resonating sounds of life.

As Debbie walks with Jessie and with several other teachers (including Dora Jilpngarri, Snowy Kulmilya, Kitty Lariyari Dadada and Hobbles Danaiyarri), she charts their extraordinarily detailed knowledge of this more-than-human world (see Chapter 3). She records and maps the zones of their walking, which animals and plants are present and which have been lost by the impacts of white settler pastoralists and those 'four-legged soldiers in the army of conquest' (Rose 2004, 86); cattle, whose hard hooves trample many edible native grasses into extinction and degrade the soil, creating eroded gullies and riverbanks. Still, Debbie takes a 'short walk with cattle' too. She reveals how the lives of the Yarralin mob have been remade with large pastoral stations like Victoria River Downs—where men were employed as stockmen, women primarily as domestic labourers in hard, corralled labour regimes, fed on meagre rations of flour, beef, sugar and tea. And when the wet made station work impossible, the mob was released

---

3   The centrality of walking in both Indigenous and settler stories about the Red Centre is explored in a book by Glenn Morrison (2017). He sees walking as a deeply cultural, even political, act and offers a comparative and cross-cultural analysis of six texts, which share a foundation in place-making through walking and writing. I thank Richard Davis for this reference.

to go 'walkabout', find bush tucker and celebrate through ceremony. This imposed colonial seasonality eclipsed Indigenous rhythms whereby the time of abundance just after the wet season was the time for large regional rituals.

But, although Indigenous Australians may have playfully seen this time of 'walkabout' as like a holiday from hard work and constant white surveillance, this was the antithesis of the white Australian myth of 'walkabout' as aimless wandering. Debbie stresses how Indigenous nomadism is a highly purposive movement across an animated Country, responding to the 'call of others', following sites of abundance as seeds, berries and fruits ripened, bush creatures like kangaroos and goannas offered good game for hunting and fish, turtles and crocodiles were surging in rivers and estuaries. Debbie dismisses the idea that Indigenous Australians, labelled as 'hunter-gatherers', were 'parasites' on an inert nature. Rather she emphasises the symbiosis of species and how humans exercised sensitive control over other beings, for instance storing seed from foraging, through fish trapping and what Rhys Jones provocatively called 'fire-stick farming' (Jones 1969).

This pervasive practice entailed a highly controlled burning of vegetation—creating open grasslands so that game could be more easily hunted, sustaining plant harvests, preserving refugia-like forests for people and animals and promoting fire-sensitive vegetation like acacia. Rainforests and sacred sites were never burnt. There was likely an aesthetic element to this burning—creating clarity and openness and an ecological mosaic. Drawing on the insights of her teacher April Bright, Debbie stressed the diversity of controlled burning regimes: hot and cold burns, in big winds and slow winds. Before lighting a fire, it was critical to know where the fire would stop. Rainfall, vegetation and landforms determined which was the best season—in the northern flood plains the appearance of yellowing grass was the signal to start burning towards the end of the wet season; in the monsoonal tropics it was from April through to June when the long grass was still green; in the desert, burning could be throughout the year, although the hot summer months were avoided (see Rose 1995). There was a responsibility to burn one's own Country, but white settlers often tried to prohibit such burning, seeing fire as a threat to themselves, cattle and station homesteads (Chapter 7, 240–41). The consequences of such prohibitions are palpable in escalating bushfires across the Australian continent (see Gammage 2012). Debbie celebrated the slow acceptance of Indigenous Australian methods of burning by Anglo-Australian land managers and this has grown since the time of her writing and especially after the uncontrolled horrific bushfires of our Black Summer of 2019–20.

Totemism: The kinship of more-than-human

Integral to such Indigenous practices of caring for Country was an affirmation of kinship which was more-than-human. This has been dubbed 'totemism'. Debbie laments the inadequacy of stock anthropological analyses of totemic creatures as 'good to eat' (Malinowski 1948) or 'good to think' (Lévi-Strauss 1963). Functionalist and structuralist interpretations alike are still predicated on western binaries of human/non-human, nature/culture. Debbie offers both a polemical critique of her anthropological ancestors and a perspective on totemism which is far more nuanced and fluid. To quote just a few words of her compelling critique:

> These questions mattered to people who believed themselves to be fundamentally different from both animals and savages. Their project had the happy benefit of refitting under a new paradigm a set of distinctions that were both foundational and self-serving. Civilisation was marked by a separation of culture from nature, so it was said, and it followed from this that a world view which posited intimate physical relations between people and animals must be understood as an absence of civilisation and must therefore constitute an evolutionary stage at which humans were not separated from nature. (Chapter 4, 150)

Fundamental to Indigenous concepts and practices of totemism is that lack of separation between nature and culture, between human and non-human. The kinship relations between humans are mirrored in the relations of plants; some are brothers, but most are cross-cousins or matrilineal kin (both called 'mates'). Debbie highlights how the Indigenous sense of ecology is one of connectivity and mutual benefit in place (Chapter 7). Her teachers especially emphasised responsible relations—that human beings are not in an extractive relation to 'resources' but have a responsibility to ensure the abundance of all other life into the future. She is critical of natural resource management (NRM) as perpetuating an extractive relation to nature as 'resources' and promoting an 'unsustainable separation of nature and culture' with disastrous long-term consequences (Chapter 8, 278). Moreover, Indigenous relations between humans and non-humans are not siloed in boxes like the bounded totemic clans of functionalist theory. They are not predicated on atomistic western thought whereby a person is a singularity, whose boundaries are coterminous with a body: 'the person achieves their maturity and integrity through relationships with people, animals, Country and Dreamings' (Chapter 4, 158). Genres of people and other creatures crosscut and overlap, emplaced both in dwelling and in movement across Country, again creating a mosaic pattern (Chapter 3, 138ff, 34, 38).

'Wild' Country and 'double death'

The idea of 'wild Country' expressed by many of Debbie's teachers, laments the loss of such responsible relations in caring for Country. Hobbles Danaiyarri observed: 'before *kartiya* [whitefellas], blackfellas bin just walking around organising the Country' (Chapter 3, 83). We can hear in this an implicit critique that whitefellas have been disorganising the Country—white settlement and dispossession has led to degradation and devastation, to the Country becoming 'wild'. Many of Debbie's teachers like Dora Jilpngarri, Doug Campbell, Riley Young and Old Jimmy talk about how cattle have caused the loss of water, with rivers receding and becoming shallower, and a host of edible plants diminishing or disappearing altogether—yams, seed-bearing plants and bulbs and parsley-like plants that grew around springs and billabongs. Bandicoots, bilbies, brush-tail possums and 'native cats' have all disappeared. These absent animals are the totemic relations of living people 'but the animals themselves, the non-human descendants of the ancestral totemic figure, are gone' (Chapter 8, 271).

Ecological devastation is graphically visible in many photos by Darrell Lewis. Massive erosion, desiccated riverbanks, rivers blocked with silt, billabongs and springs without water and lilies, soils turned into scald areas, trees struggling to survive, noxious weeds like rubber bush and devil's claw invading—we can clearly see all this. But much devastation is also invisible—the absence of previously abundant creatures, the extinction of several species—documented not just by scientific surveys but in the full-bodied memories and recollections of Debbie's teachers in Chapter 4. And as Debbie stresses, 'losses ramify destructively'. The loss of eucalypt trees is a loss of food for flying foxes and native bees, homes for arboreal marsupials and ultimately loss of food, firewood and medicine for people (Chapter 1, 46).

In dialogue with these Indigenous testimonies of loss, Debbie developed the notion of 'double death'. I quote:

> Death, as we know, occurs when the spark of life flickers out or departs. In a system of life that is seriously alive, death is worked back into the fabric of life … Double death breaks up this process,

so that death leads to more death. Loss, destruction, extinctions: double death invades systems of life, undermining their capacity to sustain the relationships that enable people, organisms, or larger systems such as Country, to be seriously alive. (Chapter 1, 49–50)

The death of a person, like Debbie's dear teacher Jessie Wirrpa, means that she returns to her Country as a nurturing spiritual presence, in a recursive looping between life and death. Once a great hunter, she is now a provider, part of the nourishing ecology of place. But double death has also come to Jessie's Country—the destruction of the capacity of life to turn death into more life. It encompasses settler violence towards Indigenous people, massacres, sometimes with entire clans eradicated, ecocidal and cultural devastation, loss of Country and its abundance, loss of languages and Indigenous knowledge. Double death puts Jessie and her kin in double jeopardy.

Absence/presence

As well as plumbing the depths of Indigenous philosophies of life and death, Debbie counterpoints and contrasts them with an array of western philosophies, including Christian and Judaic notions of the divine. In contrast to the prevalent idea of God as an unseen or absent presence, Indigenous divinity emplaced in Dreaming tracks is an immanent presence. A dominant strand in Christianity speaks of separation—Adam and Eve are banished from the Garden of Eden and from the daily presence of God. God came down to Earth again in the body of his son Jesus. He was crucified, but after resurrection his body and spirit ascended to heaven, creating for the faithful a sense of haunting loss. Thus, many Christians see this world as 'fallen', a material world of lesser value than the transcendent spiritual world. After ecofeminist philosopher Val Plumwood (1990), Debbie observes how Plato's philosophies similarly elevated the ideal world over the real, the transcendent spirit over the body. Such western theologies and philosophies are saturated with a sense of separation and of loss, and often a longing to close the ontological chasm and return to a unitary state of being. Especially in the post-Holocaust West—the 'what-is' of the world is seen as a 'broken and wounded terrain of catastrophe and risk' (Chapter 2, 72). Today, we might add that in this epoch of intensifying climate crisis, a global pandemic and renewed threats of nuclear war, dystopian visions of a damaged planet are even more pervasive.

The Indigenous notion of presence, in stark contrast to dominant western philosophies, is always emplaced and particular, located in living beings and places. To be present is to be in relation to another in intersubjectivities engaging not just humans, but more-than-human presences, other creatures and Dreamings. Dreaming ancestors created a cosmogony of movement, the traces of which guide contemporary movements. In everyday and ritual work, the flourishing of ephemeral life is ensured. The 'perduring life of creation' (Chapter 2, 73), the Dreaming, is thus actualised in the present. Debbie does not romanticise this Indigenous philosophy of divine flourishing—drought and floods, scarcity and hunger are integral presences too. But the underlying vision unites human and non-human, body and spirit, ideal and real.

A dance with time: Creation and return

In her final chapter Debbie highlights how ecological time and individual time are sequential and irreversible; a living creature comes into the world, lives and dies, children follow parents. But Indigenous concepts of the Dreaming see creation as linked to place and ongoing. 'It is not sequential, irreversible or unbounded, but rather is the continuous happening of emerging life' (Chapter 9, 283). So, when a person dies, one spirit returns to Country and takes care of the living while another flows into ecological time 'reborn either directly as a new person, or indirectly through animals to human personhood' (Chapter 9, 283). The dead keep returning to life. Again, in conversation with Val Plumwood, Debbie contrasts this with western tendencies to either resist death, see death as sacrifice or redeem life through the passage to an afterworld which is not of this world. The relation between life and death is configured as a battle where death triumphs. For Indigenous people of the Victoria River region, life not death has the last word. The bones of past people bind them to Country, and life re-emerges from Country as babies are born.

Debbie reviews the decades of anthropological debate about Indigenous understandings of human conception and discerns there a 'perverse tenacity' in such scholarship seeking to unlock secret business unavailable as public knowledge. Without trespassing into the terrain of secret knowledge, she sees the bringing forth of life as the work of both men and women in relation to sources of life in Country. 'Dreaming Women left unborn babies at various sites in their travels. These are the repositories of future generations' (Chapter 9, 289). An unborn baby seeking life leaps into an animal, the father kills that animal and gives food to the mother, who then realises she is pregnant. Baby spirits move into the nurturing womb and are born onto the ground in the mother's blood. So, Hobbles Danaiyarri was a barramundi, whom his father speared and gave to his mother to eat, so that the spirit became a baby who is now the man Hobbles. He had a small mark on his right temple where his father speared the fish. Years before, the barramundi was born from the spirit of an Aboriginal man who was killed by white men while fishing near Wave Hill (Chapter 9, 291).

Dreaming Women walked across the land, menstruated, conceived and gave birth—and their menstrual and birthing blood soaked into the land and created sites of great power and danger. Many of these sites are red ochre deposits, some belonging to women, some belonging to men, but exclusively of that Country. Debbie observes how menstrual blood in the West is often seen as a threat to the masculinist ideal of a unitary, controlled individual body—it is wet, leaky and uncontrolled and needs to be secreted or contained in the body by tampons. By contrast, in Indigenous conception women's blood is seen as sacred and potent, named and visible. Its porosity is celebrated. Since flesh is shared across species, a woman's body is shared with the bodies of other persons, other living things and Country. Blood flowing outside the body is a sign of permeability and connection, not loss. Menstrual blood is a sign of potential fertility, eliciting love and lust; unless properly managed it can make men 'wild'. Significantly, Debbie notes that in pre-colonial Australia, women experienced far fewer menstrual periods, with the late onset of menstruation and long periods of lactation and, especially in the desert and during periods of drought or famine, amenorrhoea (absence of menstrual periods) due to body fat falling below the critical threshold. This rarity perhaps intensified the potency of menstrual blood. In the past women birthed on the ground and their blood connected their person to Country. The umbilical cord and placenta connecting mother and baby was treated with care and the baby was rubbed with soil and ashes (which darkened its skin and helped resist the colonial removal of babies with mixed ancestry). The Earth itself is as a mother, its holes and caves wombs (Chapter 9, 298).

In ritual men let their own blood flow; a release of the potent blood of the Dreaming Women. Debbie criticises those analysts who see this as a symbolic appropriation of women's reproductive power, seeing men and women not in opposition but in complementary procreative relations, in shared responsibility for new life. Blood of all kinds—menstrual, birthing, arm blood, penis blood, red ochre—'is powerful and transformative'

(Chapter 9, 297). Debbie acknowledges this is dangerous territory and she stresses that such knowledge—both public and undeclared—should be treated with respect and care.

Debbie concludes her book with a reflection on how Indigenous ceremony brings Dreaming power and presence into the time, place and bodies of the performers. She is here in conversation with ethnomusicologist Cath Ellis (1984; see also Rose 2000a) who suggested that complex patterns in the timing of music mesh into a perfected totality—which lift up Dreamings from the Earth in cosmogonic action. In the intervals between there are spontaneous joking episodes. Debbie extends Ellis's analysis from song to dance—as the body connects to the Earth the Earth propels your voice into the night sky: 'You are dancing the Earth, and the Earth is dancing you' (Chapter 9, 302). As background and foreground flip in song and dance, an iridescence is created, 'an exultant awareness of life in action' (Chapter 9, 303). The bodies of the dancers are painted with earth and the designs of Country and of animals they are experiencing in daily life. The ephemeral character of everyday life and the enduring character of the potency of Dreamings in ceremony create the dance of time. 'The ephemeral calls up the enduring … in new waves of patterned life' (Chapter 9, 305).

The afterlife of *Dreaming Ecology*: A return?

The potency and poignancy of these words is amplified by the fact of Debbie's own death and the deaths of most of her teachers whom we get to know through this book and many other writings by Debbie. I now reflect on the 20 years since Debbie last worked on this manuscript. Much has happened during that time to strengthen my sense that this book was prescient.

First, I focus on how this book might be situated in the world of scholarship—not just the anthropology of Indigenous Australia but the broader terrain of ecological humanities. One of the most robust public debates in recent time has been the critique of Indigenous author Bruce Pascoe's[4] nationally acclaimed *Dark Emu* (2014) by anthropologist Peter

---

4   Bruce Pascoe has diverse Indigenous links to Yuin (NSW), Bunurong (Victoria) and Palawa (Tasmania) peoples.

Sutton and archaeologist Keryn Walshe (2021). I do not enter the details of that debate, which risked becoming hostage to polarising 'culture wars' in Australia. There have been a number of excellent contributions by Indigenous and non-Indigenous authors (e.g. Davis 2020, 2021; Griffiths 2019; Norman 2021; Rowse 2021). Pascoe graciously welcomed the book as contributing to a discussion of First Nations history (McKenna 2021b).[5]

In *Dreaming Ecology*, Debbie does not endorse the idea that Indigenous Australians were cultivators but through the knowledge and the practice of her teachers she shows the complexity of hunter-gathering life, its highly purposive character, responding to the rhythms of life and controlling and sustaining life through the controlled use of fire and 'holding' Country by ensuring the sustainability of other species in symbiotic connectivities. Her teachers' perspectives on this embraced material and spiritual dimensions in a holistic ethos of life. Yet, she does not resile from revealing the scars of the 'lash of colonisation'. She shows in graphic detail the ecological devastation consequent on white settlement—how massacres, dispossession and the arrival of cattle transformed Country; many species became rare or extinct, water became scarce and the seasonal rhythms of Indigenous life were profoundly disrupted. The commitment to the enduring power and presence of Dreamings thus emerges as a defiant resistance to how Country has become 'wild'. Moreover, she shows how Eurocentric binaries which separated nature from culture (utterly foreign to Indigenous ontology) underlay imperialist evolutionary schema which separated hunter-gatherers from cultivators on the alleged march of human progress. This difference was of course foundational for the white myth of terra nullius in Australia.

Although she did not align herself in this way, Debbie's ethnography resonates with what has been called the 'ontological turn' in anthropology, associated with the works of Roy Wagner and Marilyn Strathern in Papua New Guinea and Philippe Descola and Viveiros de Castro in the Amazon, all of which challenged western binaries between nature and culture.[6] Although some have seen this as a radical rupture in anthropology, some

---

5   Mark McKenna, historian and author of *Return to Uluru* (2021a), a searing history of frontier violence, points to the attacks both on the person of Pascoe and his book *Dark Emu* by several right-wing commentators in Australia, including Andrew Bolt on *Sky News Media*. He observed: 'The destructive grip of the culture wars—in which denigration, public shaming or abuse—makes it difficult to grasp this moment for what it truly is: an opportunity to deepen Australians' knowledge of our Indigenous cultures' (2021b). A thought-provoking documentary by the Australian Broadcasting Commission, *The Dark Emu Story*, screened from 18 July 2023, engages deeply with the debate.
6   Some of their work includes Roy Wagner (1975, 1986); Marilyn Strathern (1988); Philippe Descola (1994); Descola and Lloyd (2013); Viveiros de Castro (1992); and de Castro and Danowski (2016).

argue that this is continuous with the dormant potential in all anthropology for self-conscious reflexivity, experimentation and openness to challenge the presumptions, the ontologies of 'what is' in one's own culture and indeed the anthropological discipline (see Holbraad and Pedersen 2017).

Dreaming Ecology similarly challenges that entrenched western binary of nature and culture. Significantly, this was initiated from the 1980s and 1990s by feminist anthropologists like Carol MacCormack and Marilyn Strathern (1980) and Donna Haraway (1989, 1990, 2008) and by ecofeminist philosophers like Val Plumwood (1990, 1993), with whom Debbie sustained a vigorous dialogue. That challenge was later developed by male anthropologists like De Castro (1992), Descola (1994) and Tim Ingold (1994) and, in the history of science, by Bruno Latour (e.g. 1993, 2013). In her most recent work on the climate crisis Haraway resolutely speaks of natureculture (2016).

But Zoe Todd, Indigenous feminist scholar from Edmonton in Canada, asks whether this 'ontological turn' is just another word for colonialism. She laments how celebrated white scholars like Bruno Latour and Viveiros de Castro do not fully acknowledge their debts to Indigenous thinkers and writers, rarely citing the publications of a long list of North American Indigenous scholars she quotes. She argues:

> When we cite European thinkers who discuss the 'more-than-human' but do not discuss their Indigenous contemporaries who are writing on the exact same topics, we perpetuate the white supremacy of the academy. (Todd 2016, 18)

Moreover, she suggests this is more than just academic colonialism, but entails cherry-picking Indigenous thought without acknowledging the broader political and legal situation and the relationality of Indigenous people and scholars.

In contrast with such academic colonialism, Debbie constantly juxtaposes the ontological frictions between Indigenous and western views, an approach she dubbed 'firestick wisdom' (Rose 2011, 13–16), 'rubbing them together to see what sparks might be produced' (Van Dooren and Chrulew 2022, 3).[7] Debbie's work constantly names and acknowledges her Indigenous teachers from the Yarralin mob. And, as Linda Payi Ford observes in her Prelude,

---

7   Anna Tsing offers a rather different metaphor for this shared dialogical approach: a 'call-and-response as it refuses to stop at the lip of ontological ravines' (Tsing 2022, 16).

the book *Country of the Heart* (2002) recognised Linda, her mother, her sister and April Bright as co-authors. Debbie enthusiastically promoted Indigenous Australian scholars like Linda Payi Ford and historians Humbert Tommy Nyuwinkarri and Hobbles Danaiyarri. Moreover, she was acutely aware of the political and legal situation of her work, engaging in advocacy in arduous land claim work in 20 cases, not just in the Northern Territory as Linda describes in her Prelude, but also on the Yorta Yorta claim in Victoria and New South Wales.[8]

Dreaming Ecology needs to be situated not just in the field of anthropology but in the broader terrain of environmental humanities—a field which Debbie helped to create and sustain not just through her own writing but through the establishment first of a section in the *Australian Humanities Review* and the co-founding of the international journal, *Environmental Humanities*. This book juxtaposes Indigenous knowledge and practice with the insights of western environmental sciences, in the classification of plants and animals for instance, and celebrates the holistic approach in the ecology of the time. Since the period of her research and writing a rich corpus of transdisciplinary scholarship has emerged, often involving challenging collaborations between environmental scientists, social scientists and humanities scholars: animal and multispecies studies; philosophies of ethics and justice embracing the more-than-human and extinction studies; and importantly poetry. Debbie's writing, so sensitively attuned to language, deep storytelling and aesthetics inspired several Australian poets including Martin Harrison, Peter Minter and her later partner Peter Boyle.[9] Such transdisciplinary writing has intensified as the scale of the climate crisis has become more urgent and more acute.

---

8   This is how she expressed it:

I have worked on more than twenty land claims, native title cases, and Aboriginal land disputes, in some cases working with the claimants and in other cases working for the Aboriginal Land Commissioner. I have carried out sacred sites surveys throughout the Victoria River District. In the course of this work I have become an experienced bush woman: I have travelled by truck, foot, boat, and helicopter, have driven cross-country through sand and mud, across boulders and through the long grass, and have slept out in a wide variety of places and conditions. (Rose n.d.)

9   Martin Harrison and Debbie engaged in a dialogue where she wrestles with the relation between text and ecology (2013). Peter Boyle dedicated his book-length poem (2019) to Debbie. This was written during her terminal illness and explored shades of dark and light, grief and joy, death and life. It won the Kenneth Slessor Prize in 2020. Thanks to Richard Davis for these important observations about Debbie's broader influence.

Thom Van Dooren and Matthew Chrulew (2022) observe in their fine edited volume *Kin: Thinking with Deborah Bird Rose* that Debbie's concerns were not just scholarly. Both her life and work were

> grounded in and animated by a world of kin. This is a world of interwoven, intergenerational, more than human connectivity that both sustains and obligates, calling out for care and responsibility … the question that she returned to relentlessly—is how we are to keep faith with such a world in the midst of the ongoing processes of colonization and extinctions, of ecocide and genocide. (Van Dooren and Chrulew 2022, 2)

This expanded sense of kinship is, as they observe, thoroughly grounded in the insights of her adopted family, her friends and teachers in Yarralin and Lingara, over several decades. She worked with the Yarralin mob as with many other Indigenous people across Australia to reclaim part of the lands stolen from them. In 2016 the Traditional Owners of the Yarralin area finally gained Aboriginal freehold titles to their lands.

As I write, the lands around Victoria River are flooded in what has again been described as 'unprecedented' and hundreds of people have been evacuated to temporary shelter in Katherine and Darwin. The increased severity and frequency of natural disasters—bushfires and droughts, floods and cyclones, storm surges and sea level rise—has long been predicted by environmental scientists as the result of global heating due to the accumulation of greenhouse gases in our shared atmosphere. The ecological devastation which Debbie and Darrell witnessed and recorded from the 1980s as the consequence of dispossession by white settlers and pastoralism has been compounded by the global effects of a climate crisis emergent from a colonising capitalism and an extractive relation to Earth, ocean and atmosphere (see Ghosh 2021)—the antithesis of the Indigenous relation predicated on care and the mutual connectivity of all life.

Australia is at this moment on the cusp of a major referendum which later in 2023 will decide whether a Voice to Parliament for Indigenous Australians, as envisaged by the Uluru Statement from the Heart (2017), will be enshrined in our constitution.[10] Indigenous Australians clearly do not share the wealth or wellbeing of non-Indigenous Australians across a vast array of indicators—life expectancy, health, education, employment, housing—and experience far higher rates of domestic violence, child removal and

10   Alas this referendum was lost to a loud 'No' campaign saturated with disinformation.

incarceration (including children and youth). So far efforts to 'close the gap' have largely failed and more punitive methods like the Intervention from 2007 to 2012 have had dire consequences (Altman and Hinkson 2010). It is crucial that Indigenous people are given greater powers of self-determination and control, through embedding new mechanisms like the Voice, treaties and truth-telling. In this process, the voices of those living in remote regions like the Victoria River District need to be heard and listened to. Such voices resonate eloquently and loudly throughout *Dreaming Ecology*. We hope this book will contribute to a broader conversation in Australia as well as mark a return to the Yarralin mob—a giving back from Debbie to her teachers and also a return across generations.

Margaret Jolly
Emerita Professor
The Australian National University
Canberra
March/August 2023

References

Altman, Jon, and Melinda Hinkson, eds. 2010. *Culture Crisis: Anthropology and Politics in Aboriginal Australia.* Sydney: University of NSW Press.

Boyle, Peter. 2019. *Enfolded in the Wings of a Great Darkness.* Sydney: Vagabond Press.

Chesterman, John. 2005. *Civil Rights: How Indigenous Australians Won Formal Equality.* St Lucia: University of Queensland Press.

Davis, Richard. 2020. 'Black Agriculture, White Anger: Arguments over Aboriginal Land Use in Bruce Pascoe's Dark Emu'. Borderlands 1 (1): 57–70.

Davis, Richard. 2021. 'Review of *Farmers or Hunter-Gatherers? The* Dark Emu *Debate,* by Peter Sutton and Keryn Walshe (Melbourne University Press, 2021)'. *Arena Quarterly* 7 (September).

'Deborah Bird Rose'. 2023. *Wikipedia,* 6 May. Online: en.wikipedia.org/wiki/Deborah_Bird_Rose, accessed 17 August 2023.

De Castro, Viveiros. 1992. *From the Enemy's Point of View. Humanity and Divinity in an Amazonian Society.* Translated by Catherine P. Howard. Chicago, IL: University of Chicago Press.

De Castro, Viveiros, with Déborah Danowski. 2016. *The Ends of the World.* Cambridge: Polity Press.

Descola, Philippe. 1994. *In the Society of Nature: A Native Ecology in Amazonia.* Cambridge: Cambridge University Press.

Descola, Philippe, and Janet Lloyd. 2013. *Beyond Nature and Culture.* Chicago, IL: Chicago University Press. doi.org/10.7208/chicago/9780226145006.001.0001.

Ellis, Catherine. 1984. 'Time Consciousness of Aboriginal Performers'. In *Problems and Solutions: Occasional Essays in Musicology, Presented to Alice Moyle,* edited by Jamie C. Kassler and Jill Stubington, 149–85. Sydney: Hale & Iremonger.

Gammage, Bill. 2012. *The Biggest Estate on Earth: How Aborigines Made Australia.* Crows Nest: Allen and Unwin.

Ghosh, Amitav. 2021. *The Nutmeg's Curse: Parables for a Planet in Crisis.* London: John Murray. doi.org/10.7208/chicago/9780226815466.001.0001.

Griffiths, Tom. 2019. 'Reading Bruce Pascoe'. *Inside Story,* 26 November. Online: insidestory.org.au/reading-bruce-pascoe/, accessed 4 March 2023.

Haraway, Donna. 1989. *Primate Visions: Gender, Race and Culture in the World of Modern Science.* London: Routledge.

Haraway, Donna. 1990. *Simians, Cyborgs and Women: The Reinvention of Nature.* London: Routledge.

Haraway, Donna. 2008. *When Species Meet.* Minneapolis, MN: University of Minnesota Press.

Haraway, Donna. 2016. *Staying with the Trouble: Making Kin in the Chthulucene.* Durham, NC: Duke University Press. doi.org/10.2307/j.ctv11cw25q.

Harrison, Martin, and Deborah Bird Rose. 2013. 'Postscript. Connecting: A Dialogue between Deborah Bird Rose and Martin Harrison'. In *Writing Creates Ecology and Ecology Creates Writing,* edited by Martin Harrison, Deborah Bird Rose, Lorraine Shannon and Kim Satchell, special issue of *Text* 20 (October). Website series online: www.textjournal.com.au/speciss/issue20/Harrison&Rose_Postsc.pdf, accessed 22 August 2023.

Holbraad, Martin, and Morten Axel Pederson. 2017. *The Ontological Turn: An Anthropological Exposition.* Cambridge: Cambridge University Press. doi.org/10.1017/9781316218907.

Ingold, Tim. 1994. *From Trust to Domination: An Alternative History of Human–Animal Relations.* London: Routledge.

Jones, Rhys. 1969. 'Fire-stick Farming'. *Australian Natural History* 16: 224–28.

Kirksey, Eben. 2021. 'Obituary: Deborah Bird Rose (1946–2018)'. *Asia Pacific Journal of Anthropology* 22 (1): 81–83. doi.org/10.1080/14442213.2020.186 7956.

Latour, Bruno. 1993. *We Have Never Been Modern.* Translated by Catherine Porter. Cambridge, MA: Harvard University Press.

Latour, Bruno. 2013. *An Inquiry into Modes of Existence: An Anthropology of the Moderns.* Cambridge, MA: Harvard University Press.

Lévi-Strauss, Claude. 1963. *Totemism.* Boston, MA: Beacon Press.

Lewis, Darrell. 1988. *The Rock Paintings of Arnhem Land, Australia. Social, Ecological and Material Culture Change in the Post-Glacial Period.* British Archaeological Reports No 415.

Lewis, Darrell. 1996. *The Boab Belt: A Survey of Historic Sites in the North-Central Victoria River District.* Report prepared for the Australian National Trust (NT).

Lewis, Darrell. 2002. *Slower than the Eye Can See: Environmental Change in North Australia's Cattle Lands. A Case Study from the Victoria River District, Northern Territory.* Darwin: Tropical Savannas CRC.

Lewis, Darrell. 2007. *The Murranji Track: 'Ghost Road of the Drovers'.* Rockhampton: Central Queensland University Press.

Lewis, Darrell. 2012. *A Wild History: Life and Death on the Victoria River Frontier.* Clayton: Monash University Publishing.

Lewis, Darrell, and Charlie Schulz. 1995. *Beyond the Big Run: Station Life in Australia's Last Frontier.* St Lucia: University of Queensland Press.

Lewis, Darrell, and Deborah Bird Rose. 1988. *The Shape of the Dreaming: The Cultural Significance of Victoria River Rock Art.* Canberra: Australian Institute of Aboriginal Studies.

MacCormack, Carol P., and Marilyn Strathern, eds. 1980. *Nature, Culture and Gender.* Cambridge: Cambridge University Press.

Malinowski, Bronisław. 1948. *Magic, Science and Religion, and other Essays.* London: Souvenir Press.

McKenna, Mark. 2021a. *Return to Uluru.* Carlton: Black Inc.

McKenna, Mark. 2021b. 'Bruce Pascoe Has Welcomed the *Dark Emu* Debate – And so Should Australia'. *Guardian*, 25 June. Online: www.theguardian.com/commentisfree/2021/jun/25/bruce-pascoe-has-welcomed-the-dark-emu-debate-and-so-should-australia, accessed 4 March 2023.

Morrison, Glenn. 2017. *Writing Home: Walking Literature and Belonging in Australia's Red Centre*. Melbourne: Melbourne University Press. doi.org/10.2307/jj.5993303.

Muecke, Stephen. 2020. 'Her Biography: Deborah Bird Rose'. *a/b: Auto/Biography Studies* 35 (1): 273–77. doi.org/10.1080/08989575.2020.1720202.

Norman, Heidi. 2021. 'How the Dark Emu Debate Limits Representations of Aboriginal People in Australia'. *The Conversation*, 8 July. Online: theconversation.com/how-the-dark-emu-debate-limits-representation-of-aboriginal-people-in-australia-163006, accessed 4 March 2023.

Pascoe, Bruce. 2014. *Dark Emu: Black Seeds: Agriculture or Accident?* Broome: Magabala Books.

Plumwood, Val. 1990. 'Plato and the Bush: Philosophy and the Environment in Australia'. *Meanjin* 49 (3): 524–36.

Plumwood, Val. 1993. *Feminism and the Mastery of Nature*. London: Routledge.

Rose, Deborah Bird. 1991. *Hidden Histories: Black Stories from Victoria River Downs, Humbert River, and Wave Hill Stations, North Australia*. Canberra: Aboriginal Studies Press.

Rose, Deborah Bird. 1992. *Dingo Makes Us Human: Life and Land in an Australian Aboriginal Culture*. Cambridge: Cambridge University Press.

Rose, Deborah Bird, ed. 1995. *Country in Flames: Proceedings of the 1994 Symposium on Biodiversity and Fire in North Australia*. Canberra: Biodiversity Unit, Department of the Environment, Sport and Territories; Darwin: The North Australia Research Unit, ANU. Online: openresearch-repository.anu.edu.au/bitstream/1885/282735/1/b18944231.pdf, accessed 31 January 2023.

Rose, Deborah Bird. 1996. *Nourishing Terrains: Australian Aboriginal Views of Landscape and Wilderness*. Canberra: Australian Heritage Commission. Online: www.ceosand.catholic.edu.au/catholicidentity/index.php/sustainability/sustainability-and-aboriginal-education/91-nourishing-terrains/file, accessed 30 January 2023.

Rose, Deborah Bird. 1997. 'Common Property Regimes in Aboriginal Australia: Totemism Revisited'. In *The Governance of Common Property in the Pacific Region,* edited by Peter Larmour, 127–43. Canberra: National Centre for Development Studies.

Rose, Deborah Bird. 1998. 'Consciousness and Responsibility in an Australian Aboriginal Religion'. *Nelen Yubu.* Dickson, ACT: Daramalan College. Online: misacor.org.au/images/Documents/NelenYubu/NY23.PDF, accessed 30 July 2023.

Rose, Deborah Bird. 2000a. 'To Dance with Time: A Victoria River Aboriginal Study'. *Australian Journal of Anthropology* 11 (2): 287–96. doi.org/10.1111/j.1835-9310.2000.tb00044.x.

Rose, Deborah Bird. 2000b. 'Love and Reconciliation in the Forest: A Study in Decolonisation'. Hawke Institute Working Papers No 19. Adelaide: University of South Australia.

Rose, Deborah Bird. 2004. *Reports from a Wild Country: Ethics for Decolonisation.* Sydney: UNSW Press.

Rose, Deborah Bird. 2011. *Wild Dog Dreaming: Love and Extinction.* Charlottesville, VA: University of Virginia Press.

Rose, Deborah Bird. 2022. *Shimmer: Flying Fox Exuberance in Worlds of Peril.* Edinburgh: Edinburgh University Press. doi.org/10.1515/9781474490412.

Rose, Deborah. n.d. 'Curriculum Vitae: Deborah Bird Rose'. University of New South Wales. Online: unsw.academia.edu/DeborahRose/CurriculumVitae, accessed 22 August 2023.

Rose, Deborah Bird, with Sharon D'Amico, Nancy Daiyi, Kathy Deveraux, Margy Daiyi, Linda Ford and April Bright. 2002. *Country of the Heart—An Indigenous Australian Homeland.* Canberra: Aboriginal Studies Press.

Rowse, Tim. 2021. 'The Teller and the Tale'. *Inside Story,* 16 June. Online: insidestory.org.au/the-teller-and-the-tale/, accessed 4 March 2023.

Strathern, Marilyn. 1988. *The Gender of the Gift: Problems with Women and Problems with Society in Melanesia.* Berkeley, CA: California University Press. doi.org/10.1525/california/9780520064232.001.0001.

Sutton, Peter, and Keryn Walshe. 2021. *Farmers or Hunter-Gatherers? The* Dark Emu *Debate.* Melbourne: Melbourne University Press. doi.org/10.2307/jj.1176863.

Todd, Zoe. 2016. 'An Indigenous Feminist's Take on the Ontological Turn: "Ontology" Is Just Another Word for Colonialism'. *Journal of Historical Sociology* 29 (1): 4–22. doi.org/10.1111/johs.12124.

Tsing, Anna Lowenhaupt. 2022. 'The Sociality of Birds: Reflections on Ontological Side Effects'. In *Kin: Thinking with Deborah Bird Rose*, edited by Thom Van Dooren and Matthew Chrulew, 15–32. Durham, NC: Duke University Press. doi.org/10.1215/9781478022664-002.

Uluru Statement from the Heart. 2017. Online: ulurustatement.org/the-statement/view-the-statement/, accessed 10 March 2023.

Van Dooren, Thom, and Matthew Chrulew. 2022. 'Worlds of Kin: An Introduction'. In *Kin: Thinking with Deborah Bird Rose*, edited by Thom Van Dooren and Matthew Chrulew, 1–14. Durham, NC: Duke University Press. doi.org/10.1215/9781478022664-001.

Wagner, Roy. 1975. *The Invention of Culture*. Englewood Cliffs, NJ: Prentice-Hall.

Wagner, Roy. 1986. *Symbols That Stand for Themselves*. Chicago, IL: University of Chicago Press.

1

Coming into Country

With Jessie

Figure 1.1. Jessie Wirrpa and the author, out bush on Victoria River Downs (VRD), 1982.

Source: Photograph by Darrell Lewis.

When Jessie Wirrpa took me walkabout she called out to her ancestors (Figure 1.1). She told them who we were and what we were doing, and she asked them to help us. 'Give us fish,' she would call out, 'the children are hungry.' When she was walking through Country she was always with a group, and that group included the dead as well as the living. As her brother Allan Young Najukpayi said:

> At night, camping out, we talk and those [dead] people [and they] listen … When we're walking, we're together. We got dead body there behind to help … Even if you're far away in a different Country, you still call out to mother and father, and they can help you for dangerous place. And for tucker they can help you. (Quoted in Rose 1992, 73)

Every place we visited was part of some story or other. Here the owlet nightjar (Jessie's and Allan's 'totem') burnt his whiskers, there he flew away. Here the bandicoot and possum were hanging out, and over there the flying foxes chucked their spears. Jessie spoke to the Dreamings too, calling out to them to let them know that the people who were there belonged there, that strangers were accompanied, that this was all lawful.

As we walked, we took notice of other living things. When we startled the cockatoos so that they squawked and flew away Jessie laughed because they were making a fuss about nothing. When the green flies bit us, we knew the crocs were laying their eggs, and Jessie began to think about going walkabout to those places. When the *jangarla* tree (*Sesbania formosa*) started to flower, we knew, or Jessie knew, that the barramundi would be biting. The world was always talking about itself, and Jessie was a skilled listener and observer. She knew how to interpret tracks, too. She knew who made what track, and she knew when and why those tracks were made, and where the animal was likely to be now.

We never walked aimlessly, but always opportunistically. Jessie knew the soils, creeks, ridges, springs, shelter and other aspects of her Country. Each such niche was a habitat: this is where we go for conkerberries, this is where we go for lilies, and here is where we'll spot the signs of yams. When we went for conkerberries she knew it was time to go because the fireflies had appeared. We saw the bushes from a distance, and she pointed to where a turkey had been eating the berries. When it is eaten up that high, it's a turkey; higher again and it's an emu. Then she hushed us, because she saw marks of a goanna under the bush. With a few hand signs she organised the kids to circle the conkerberry bushes, and told the young women to flush out, stun and kill the goanna.

Every species of tree had a name, and every tree lived within relationships of beneficial connections. This one provided shade for humans, and food for black cockatoos; that one you could make medicine from; another one had bark that you would turn into ashes to roll your chewing tobacco in.

Many of the species of grass, and most of the shrubs and other plants, had names and were in relationships. There were stories for them too: the owlet nightjar didn't crack his lily corms properly and that is why they exploded in his face and burnt his whiskers. The bandicoot and possum were digging 'cheeky yams' on that hill—that's where they grow.

Everywhere Jessie went she encountered signs of former activities. A discarded stone spear point, some charred sticks from a campfire, a Dreaming tree that got knocked by lightning when her oldest father died. No distinction between history and pre-history for Jessie; in her Country the present rolled into the past on waves of generations who all belonged there. As she walked, she told the stories: here the owlet nightjar burnt his whiskers and flew away, there we saw the goanna track; here he jumped out, and over there Margaret hit him. Debbie didn't know what to do, and we all had a good laugh. Here we cooked him—'good dinner camp, that one.'

I know that Jessie's Country gave her life; I walked with her, and it gave me life too. It nourished her, and she took care of it. She was a presence in her Country, and her Country knew her; Country fed her and her group; it held the signs of their lives and stories and continued to bring forth life. Tagging along behind her, I did my best to learn.

The questions I brought with me from America in 1980 were philosophical in the main. I wanted to learn something about the meaning of life and death from the viewpoint of some Australian Aboriginal people. These first questions went onto the back burner while I threw myself into participant observation. I had a lot to learn about how to make a fire, how to catch a fish, how to understand North Australian Kriol and other languages. Of course, I had to learn who people were, and this meant learning about their connections—to other people, other plants and animals, to Countries and to Law. That meant learning terms for species, and learning placenames, Dreaming tracks, places of danger. I followed behind Jessie as I learned how to behave properly in the face of all these people, animals, plants and places. By the time I returned to my research questions I was in a different world, a living world made up of Countries. I had come to understand Country as a geographical area and as a system of nourishing life, a place within which responsibilities and reciprocities are recursive, and in which living things take care of their own (Rose 1992, 1996).

Country is full of sentience—animals, many plants, Dreamings, the ancestors and other things like hills or stones take notice, as people say. Jessie took notice too, and she knew that all these other beings were taking notice of her. Her footprints, her fires, her songs and stories, her visits and her calling to the Country were all communicative acts. They worked within the broader communicative system of Country, and they intended to be noticed. Correspondingly, her way of teaching required one to pay close attention. When I hollered for help and she showed me what to do she would say: 'I'll show you once, after that you do it yourself.'

Jessie took care of her Country, and her Country took care of her. Country was the ground of her being and she herself was a walking and ephemeral nexus of Country, Dreamings and care. Nothing stood alone in Jessie's Country; everything was happening because of the care of others. She took notice, and I never knew her to hear silence. Her attending presence was, to her knowledge, always reciprocated. Her voice and action, footprints and other traces, communicated to others. She never thought there was nobody out there.

I do not want to suggest that Jessie Wirrpa somehow escaped the lash of colonisation. It descended upon her in the form of conquest, and it killed most of her people (Rose 1991). Australia is a modern settler society, a nation designed to achieve the great prospects of modernity. The Enlightenment vision of liberty and equality had a rough ride in Australia; the penal colony was founded in violence, control, displacement and hierarchy. The Enlightenment vision, however, has been a powerful force in the making of a modern plural society, and in achieving justice for the marginalised and dispossessed. Along with visions of a better society, there were and are the social and cultural practices that Scott (1998, 4) defines as high modernism: self-confidence about scientific and technical progress, expansion of production, mastery of nature and rational design of social order. To these I would add a triumphal narrative of how these dreams of progress and control are being implemented for the betterment of society.

When the goal of mastery of nature and of society is coupled with 'new world' settler colonialism in contexts defined as frontiers, the result is massive eradication—of human societies and local knowledge as well as of ecosystems in the integrity of their diversity and sustainability. Faith in progress works in dynamic feedback with violence: conquest requires violence, violence leads to less diversity, less diversity (human and environmental) leads to greater demands for scientific and technological control, which leads to more intensification in favour of greater perceived productivity, which leads to less diversity. The cycle is fuelled by faith along with violence.

Jessie's Country is certainly not a place that has no knowledge or experience of modernity. Rather, she, her people and her Country know modernity all too well, for they have borne the brunt of its underside. Colonisation came late to the backwaters of the Victoria River District. Here in the savanna regions of the monsoonal north, white settlers established huge cattle stations just over one hundred years ago. Overrunning the homes of the Indigenous hunter-gatherer peoples of the region, they first shot and hunted away the local peoples, and later pressed them into service on the cattle stations as an unfree and unpaid labour force. In the mid-1960s many of the Aboriginal people in this region went on strike against the appalling system of oppression which ruled their lives, and as support for them was manifested throughout the nation, they began to articulate what to them had been an underlying purpose in all their forms of resistance: to regain control over at least some portions of their traditional homelands. By the 1990s, legal recognition of land rights had benefited some far more than others, but fuller citizenship (realised around 1967[1]) has enabled people to participate in national and international struggles for equality. The story is by no means over. Indigenous people's cultural survival continues to be contested locally and nationally.

Jessie lived for most of the twentieth century, and her life took this shape: under the name of progress, settler Australians used up her labour for much of her life (see Rose 1991). In the dry season the cattle were mustered, branded, sorted and driven to market. Aboriginal workers were required both in the stock camps (mostly men) and as domestics (mostly women). In the wet season cattle work was discontinued and station life was reduced to the basics. Aboriginal workers were sent walkabout, to travel the Country and to survive using their Indigenous knowledge and skills. Until about

---

1   According to Tim Rowse, it is a common misconception that full citizenship was granted by the 1967 referendum. The successful 1967 referendum meant that Indigenous Australians were counted in the Census and enabled the Commonwealth to pass laws affecting them. In fact, Indigenous Australians became citizens through the *Nationality and Citizenship Act 1948*—a law declaring that persons living in Australia who were 'British subjects' (as Indigenous Australians were) henceforth were citizens of Australia. However, State and Territory laws in five of the six states (not Tasmania) and in the Northern Territory continued to regulate Indigenous Australians in many ways, curtailing freedoms that we usually associate with 'citizenship'. These restrictive regulations and policies of assimilation, such as child removal, were slowly repealed, so that the cumulative effect of these changes amounted to a fuller sense of Indigenous citizenship by 1967. For more detail see Chesterman (2005)—eds.

1965 people worked for white station owners for most of the year and lived by and for their traditions during part of the year. As I discuss in subsequent chapters, people's lives in the bush were more complex than this seasonal schema suggests, but the broad outline shows that knowledge of the nomadic life, and the detailed knowledge of Country were essential skills for physical and cultural survival. At the same time, however, as I will also discuss, the impacts of cattle monoculture diminished the availability of subsistence resources to a high degree. Even before Jessie was born in 1917, large portions of the original resource base were so severely damaged as to be non-viable for subsistence; many of the springs and billabongs had dried up (or were drying up) and the predictability of resources was diminished. Pastoralism had also eradicated some of the species with which she and her family were intimately connected and on which they depended. The years since then have intensified these processes of loss. Just as living things are connected through benefit, so losses ramify destructively. The loss of eucalyptus trees, for example, is a loss of food for flying foxes and for native bees, and loss of habitat for arboreal marsupials. Such losses diminish human food supplies, as well as reduce available firewood, medicine and other benefits.

In the vast pastoral leases of the Victoria River valley, introduced pastures, thousands of miles of fences, the suppression of Indigenous fire regimes, and many other factors aimed toward improving cattle productivity have massively degraded the rangelands. As a result of the *Aboriginal Land Rights (Northern Territory) Act 1976,* there are now Aboriginal as well as settler pastoralists; all are under considerable pressure to maintain the productivity of their businesses. Pastoralists in the Victoria River District achieve economic independence through intensification of their use of the ecosystems, including ecological services, upon which their enterprises depend. The industry continues to call for more scientific knowledge to enable people to intensify the outputs of their overdeveloped monocultures.

Jessie and her mob are not independent pastoralists; the amount of land they own under Anglo-Australian law is too small to be economically viable. And yet for them the pressures are much the same: how to achieve some form of sustainability in landscapes in which the former subsistence practices are no longer viable, and in regions remote from markets.

Experience

In the dry season out here in the tropical savannas you wake up hoping somebody else has already gotten up and started the fire. The last nightjars call out, the sky starts to lighten, and you wiggle further into your swag for warmth. A big fire, a cup of tea and some sunlight change your perspective. The air is so clear, the sky so hugely blue, the light so crisp that you are eager for whatever the day may bring. By midday the world is hot again, except on those rare days when the southerly busters break through to the tropics. Then the winds fly up out of the south, from Antarctica it feels, demolishing every bit of warmth. They rush across the Central Australian deserts losing little of their cold energy, and when they hit you, you huddle up, find a windbreak and wait it out.

Such days are rare. For the most part the dry season is a time of warmth and clarity. The breezes carry the invigorating smells of spinifex and eucalyptus blossoms, both of which get you thinking of native honey. As the dry season continues past the winter equinox, the days become hotter, the nights less severely cold. The ground starts to warm up too. Previously, you thought longingly of socks, now you're happy in thongs [flipflops]. Before long you will bless those thongs because the ground has become so hot that your bare feet can't take it. Aboriginal people are happy for thongs, too. Those beautiful tough nubbly feet that have walked the Country for years and years, even they find the hot ground of the hottest time of year hard to take.

As the season progresses, the cattle eat most of the grass. What is left is dry and brittle, poor in nutrition, with little by way of substance or structure. White settlers drove the shorthorn cattle into this Country, coming out of Queensland where they had experience of hot arid Country. In the 1950s a few brahmins were introduced, and since the 1970s they predominate. Their silver hides, dignified humps and sedate faces seem part of the place. In the late dry they will cluster at the bores, losing condition and looking more stoic than elegant. The trees lose their vivid colours, except for some of the sturdy hot weather trees whose deepening colours tell you that the season is progressing. With the ground cover drying out, the soil becomes ever more friable. The ants are busy, as usual, but even they look hot and dusty.

In the late dry the humidity builds up. The air becomes hot and feels even hotter. It is exciting, too. Sometimes you see a huge red cloud that fills one whole section of the sky and comes towering toward you. It is a wind front that has picked up the red dirt. The air is filled with sticks, leaves, pieces of wood, perhaps even corrugated iron, as well as the choking dust. Everybody runs for shelter—it hits like a train, slams through your area picking laundry up off the wire along with anything else that is detachable and dropping bits of debris. Once it is gone you look around and laugh together with relief.

The first rains seem to take forever to arrive. You have watched the thunderheads move across the sky. The lightning has been fantastic in its varied forms and its prolific intensity. You are reminded of the fact that the lightning strike rate is one of the highest in the world. Some nights when you sit longing for a breath of air and watching the lightning circle around, you see off in the distance the glowing line of a bushfire, another lightning strike. Wherever you are, it seems that the rain always falls somewhere else first. It may tease you with tormenting little sprinkles, but finally it comes pouring down in glorious cool floods of relief.

The rainbows are multiple and vivid. I think of God's promise to Noah and wonder if a holy rainbow could possibly be as large, pure and vibrant as these. There are light rains and dark rains and, every once in a while, a rain unlike any other. The drops are huge and widely spaced. When the sun is shining, each drop catches the light and shines like a crystal as it slowly makes its way toward the ground. The depth of vision—bronze earth, golden grass and silver rain, is transporting; all is in perfect clarity. After rain the colours intensify. White trees turn silver, yellow grass turns gold, and the light seems to hang in the air as if it were buoyant. As the sun goes down it all becomes more intensely silver and gold until it shades into burnished copper and then goes out, leaving you breathless.

By late dry, the rivers have stopped running and are shrunk back to their permanent waterholes. Every day they seem smaller, every day more suspended in anticipation. When you go fishing you smell the flying foxes who have come to the river for the flowering eucalypts of the late dry season. The first big rains hit the desiccated and barren soil and run off into the creeks and rivers. They pull the heat out of the ground and deposit it in the water. To a North American like me, the sound of water rippling across the stony riverbed is a call to refresh myself but, when I leap in, I find it is too hot to be enjoyable. This is the run-off that stuns the fish and brings them floating and flopping to the surface.

Some years the rains settle in and hang around for weeks as cyclonic depressions move across North Australia. Other years it hardly rains at all, and the heat and humidity just go on and on without respite. Unless there is a near absence of rain (and this happens, although I have not had the opportunity to see it), new growth starts to come forth. Green grass springs up out of the ground, new leaves appear on the trees, the vines spring to life and wind and tangle along the ground and up the shrubs. The rivers are flowing well; in good years they run a banker [river flooded to the top of its banks].

The next lot of wind starts to clear away the rains. In the early dry the grasses brown off and many fruits, bulbs and tubers are ripe. The humidity departs, the nights become cooler, and as the ground dries out you can start to drive around the country again. Everything that moves is walking around now—the kangaroos, wallabies and other marsupials are on the hop, the birds are active, and the ants are in heaven. Cattle get around the country in smaller groups, moving gently along the horizon. You smell them from afar, a dense, clean herbivore smell, and you look across the country for them, to see what sort of group is there. The flying foxes are gone, but they will return; that is their way.

Life and death

Through walking with Jessie and many others, and through being taught by a large number of excellent teachers, I developed a heightened sense of life and a stronger sense of the possibilities surrounding death. These two themes run through this book, sometimes explicitly, often latently.

The first is a state of becoming; I call it 'seriously alive'. To be seriously alive is to be in connection—caring and being cared for. In Country, as I have briefly mentioned, almost everything lives within relationships of beneficial connection (Chapter 7). To be seriously alive, then, is to be in connection. Not only does the person, or organism, or Country, live from one moment to the next; they also contribute to the lives of others and gain benefit from the lives of others. To be seriously alive is to be complexly connected in relationships that sustain systems of life.

The second theme is that of ramifying death: 'double death' in my terminology. Death, as we know, occurs when the spark of life flickers out or departs. In a system of life that is seriously alive, death is worked back into the fabric of life (Chapter 9). Double death breaks up this process, so that death leads to more death. Loss, destruction, extinctions: double death invades systems of life, undermining their capacity to sustain the relationships that enable people, organisms, or larger systems such as Country to be seriously alive (Chapter 8).

Jessie's own life and death offers an excellent exposition of both concepts. As I discuss in *Dingo Makes Us Human* (Rose 1992), Yarralin people talked about the components of the human person in ways that suggested at least two animating spirits (and I also expressed caution about the use of the term 'spirit'). One of them keeps returning from life to life, so that death becomes an interval leading into transformation into new life. Often the genealogy of 'spirit' includes animals as well as humans—this life force moves through life forms and continues to bind death and life into the ongoing and emerging life of the place. Another 'spirit' returns to the Country to become a nurturing presence. These are the 'dead bodies' I spoke of earlier—the dead Countrymen (the term includes both women and men) who continue to live in Country and to whom people appeal when they go hunting. This ecology of emerging life sets up recursive looping between life and death; Country holds both, needs both and, most importantly, keeps returning death into life. The return is what holds motion in place, and in the dynamics of life and death, life is held in place because death is returned into place to emerge as more life.

Double death breaks up this dynamic, place-based recursivity. The first death is ordinary death; the second death is destruction of the capacity of life to transform death into more life. In the context of colonisation, double death involves both the death that was so wantonly inflicted upon people, and the further obliterations from which it may not be possible for death to be transformed. Languages obliterated and maybe gone forever, and clans or tribes eradicated and maybe gone forever are examples of double death.

Ecological violence performs much the same forms of obliteration. Thus, species are rendered extinct, billabongs and springs are emptied of water, and soils are turned into scald areas. This violence produces vast expanses where life founders. It amplifies death not only by killing pieces of living systems, but by diminishing the capacity of living systems to repair themselves, to return death back into life. What can a living system do if huge parts of it are exterminated? Where are the thresholds beyond which death takes over from life, and have we exceeded those thresholds violently and massively in the conjoined process of conquest and development? These processes are not always irreversible, but in many areas the answer is yes.

In Jessie's way of life and death, she has joined the other 'dead bodies' in her Country, and, like them, is becoming part of the nourishing ecology of the place. In life she was a great hunter, in death she joins the providers. Double death puts her in double jeopardy. The rivers are rapidly deteriorating from erosion and siltation, and even more severely at this time, from invasions of noxious weeds. It is probable that in the near future riverine ecologies will collapse, and with that collapse the possibility for the living to go fishing and feed their families will be radically impaired, if not completely obliterated. For Jessie, then, there's a doubling up: first her own death as a living person, and subsequently, her obliteration as a nurturer within a flourishing Country.

Ecologies

I believe that Aboriginal people's culture of connectivity has strong points of connection with contemporary ecological thought. The demands that the new ecology puts on western scholarship are a partial guide to some of the demands that Dreaming ecology puts on western scholarship. For a scholar such as myself, one of the great obstacles to engaging with Aboriginal culture is the west's atomistic legacy, a legacy that has marginalised motion, process, recursivity and connection. Gilles Deleuze, in particular, has carried the argument against atomism in social theory. His work bends western philosophy towards its own margins and invigorates history with critique arising from concepts of fluidity, multiplicity and difference (see also Douglass 1992). His work is part of the emergence of parallel domains of converging scholarship that marked the latter half of the twentieth century. On the one hand cybernetics, systems theory, theoretical physics and ecology have started to reshape how we think about the world and to give us a language for engaging with a post-atomistic cosmology (Mathews 1991). On the other hand, feminist critique, post-structural theory and philosophies of intersubjectivity are reshaping and reimagining how we think about the world and our own analytic practice. As an anthropologist I am strongly influenced by Gregory Bateson who, in his enormously influential life, contributed to almost all of the domains mentioned above.

In brief summary, the new ecology starts with the assertion that the fundamental condition of life is connection (Bateson 1972, 436; Harries-Jones 1995, 66; Mathews 1991). The implication is that being is inherently, inescapably and necessarily relational. This new thinking is part of the radically unsettling paradigm shift now taking place in the western world.

From Newton to Einstein, life was conceptualised in the mode of atoms—single entities whose essential characteristic was separation. Atomistic theory had difficulty accounting for motion; atoms were assumed to be inherently in stasis (Mathews 1991). In the twentieth century the theories of relativity, the development of quantum theory and other world view–altering achievements require us to rethink the nature of life. Relationships, not atoms, are the foundation. Motion, not stasis, characterises the universe. According to Ilya Prigogine, the shift is 'from substance to relation, to communication, to time' (quoted in Midgley 1992, 40).

Amongst ecologists, whose training is principally in the fields of science, the shifts in thinking are revolutionary. An ontology of connectivity entails mutual causality: organism and environment modify each other. Relations between organism and environment are recursive, meaning that 'events continually enter into, become entangled with and then re-enter the universe they describe' (Harries-Jones 1995, 3). Once, scientists could imagine that proper methods were capable of bringing the universe into the mind of humanity. Some theoretical physicists still hold this to be the case, but most ecologists have become more humble. Frank Egler (1977) is reported to have said that 'ecosystems may not only be more complex than we think, they may be more complex than we *can* think' (quoted in Dietrich 1992, 110). This view represents a fundamental shift from the proposition that incomplete knowledge is an obstacle to be overcome, to the proposition that incomplete knowledge is a condition of any participant in a living system.

From my perspective, the key points of connectivity are:

1. the fundamental condition of life is connection (not separation),
2. the basic unit of survival is 'organism and environment',
3. systems are holistic; knowledge is therefore of necessity partial.

I am not, of course, suggesting that Australian Aboriginal culture is either post- or pre-Newtonian/Cartesian; I would actively resist the universalising tendency implicit in forcing another culture into a position defined by western history. In writing this book I work toward encounters. I do not argue against anthropological or other theory as a primary purpose. Rather, I bring different discourses into proximity so that they may resonate with each other. The book is a site that must of course include both author and reader in encounter. I am aiming for an intersubjective process in which knowledge can arise in excess of the information and analysis presented in the text.

Care of Country

Australia is the world's most arid inhabited continent, a land of many deserts and few rainforests. It is a land of monsoonal tropics in the far north, and temperate Mediterranean climates in the coastal south. Between these two relatively well-watered areas there is a vast zone of arid and semi-arid unpredictability within which is located the greatest portion (80 per cent) of the land mass.

Before European colonisation, the only domesticated animal was a recent arrival, the dingo (*Canis lupus dingo*), who has been in Australia for only a few thousand years. During most of the period we would identify as the time of fully modern humans, Aboriginal people lived in Australia without domesticated plants or animals. They learned to know and work with their island continent, embedding themselves ever more intricately into the places which they shaped and which shaped them. It was a continent of hunter-gatherers, as Harry Lourandos (1997) puts it. Here humans developed, to the greatest degree knowable in the history of our species, a way of life that works with the grain of nature, seeking connectivities that enmesh humans ever more recursively into the places that sustain them.

Western thought pervasively and profoundly has contrasted those who cultivate the soil with those who do not. Many of the differences between cultivation and hunter-gatherer modes of subsistence have been conceptualised in western thought by reference to human intentional action in the world, and rest on the western nature–culture dichotomy (see Plumwood 1990, 1993 and other authors). According to Tim Ingold (2000, 58):

> The producer is conceived to intervene in natural processes, from a position at least partially outside it; the forager, by contrast, is supposed never to have extricated him- or herself from nature in the first place.

The nature–culture dichotomy, although now massively destabilised through feminist critique, continues to situate hunter-gatherer people ambiguously. Ingold (2000, 58) notes that contemporary usage which replaces the term 'hunter-gatherer' with the term 'forager' perpetuates both the dichotomy and the slippage: like animals, foragers graze across a landscape.

A growing body of knowledge in Australia is currently reconfiguring most of what once stood as conventional wisdom concerning Aboriginal hunter-gatherers, leading to a quiet revolution in anthropological thought concerning the relationships between Aborigines and their Country. In recent years, issues of Indigenous land management have come to be understood as questions for research. In Australia the long-term lack of research, like the lack of general public awareness of these issues, is connected with the settler view that Aboriginal people were parasites on nature. A.P. Elkin (1954, 14) gave the mark of scientific authority to this view in a book first published in 1938: 'The food-gathering life is parasitical; the Aborigines are absolutely dependent on what nature produces without any practical assistance on their part.' This view of parasitism was intricately connected to the view of *terra nullius:* the idea that the land was untransformed underpinned the idea that the land was unowned. By this logic, Aboriginal 'parasites' were excluded from forms of ownership by reason of their own nature (i.e. lack of culture).

An important corollary was that hunter-gatherers did not shape the landscape, or that the landscape was shaped by them only as a by-product of their foraging activities. Nancy Williams and Eugene Hunn's (1982) publication *Resource Managers* marks a key moment in shifting the accepted conventions surrounding these issues (see also Williams and Baines 1993). The start of the demolition of the parasite view, however, dates to Rhys Jones's (1969) work on the use of fire in a system of land management. He called this system fire-stick farming, and his use of the term 'farming' was deliberate (Jones 1995, 14). Inaccurate as it is in attributing the culture of cultivation to Aboriginal people, it provocatively targeted an intellectual and political nerve. If Aboriginal people were fire-stick farmers they could not be mere parasites, and their labour would have to be construed as productive. Within Lockean logic, having invested labour in land, there must exist the basis for land ownership.

Since Jones's original work, there have been numerous studies that show Aboriginal people's proactive care of Australian fauna, flora and ecosystems. Increasingly it is becoming evident that both the distribution and the diversity of Australian biota across the continent are artefacts of Aboriginal people's intentional actions. This is not to say that Aboriginal people have always and only managed ecosystems well; knowledge and practice are not always in synchrony for Aboriginal people, any more than for others (Lewis 1993). On the other hand, 60,000 or more years of stable life in the world's most arid inhabited continent speak to successful interactions with ecosystems. In this context we should bear in mind that Aboriginal people do not hold labour to be a sign or proof of ownership; rather, they hold knowledge to constitute such a proof.

The implications of this new knowledge of proactive care are enormous. Research is in a very early stage; it is interdisciplinary and has yet to be fully accepted within any mainstream discipline. Although there is a vestigial debate about whether Aboriginal people actually did engage in fire-stick farming (Horton 1982, 2000); today it is almost universally accepted that they consciously managed large portions of the continent through the use of fire, and contemporary Anglo-Australian land managers now seek to use fire to manage landscapes in many parts of Australia and to work with and learn from traditional owners (Head 1994; Pyne 1991).

Looking at the continent as a whole, it is now evident that Indigenous people's actions are almost certainly responsible for maintaining the open grasslands that covered much of the continent (Jones 1969), for the preservation of specific stands of fire-sensitive vegetation such as acacia (Kimber 1983), cypress (Bowman 1995; Bowman and Panton 1993) and remnant rainforests (Russell-Smith and Bowman 1992), for the protection of refugia including breeding sanctuaries (Newsome 1980), and the preservation of sources of permanent water in arid environments (Latz 1995). In addition, their actions are directly responsible for the distribution of many plants (Hynes and Chase 1982; Kimber 1976; Kimber and Smith 1987, 233), and probably for the distribution of some fauna such as freshwater crayfish (Horwitz and Knott 1995). If research continues to produce new knowledge at this current rate, it is probable that today's knowledge is only the tip of the iceberg (more extensive summaries of many of these issues are found in Rose 1996).

The idea that Aboriginal people were parasites depended on three main assumptions. First, it posits a separation between humanity and nature. Second, it assumes a linear flow of energy from the natural world to the human world. Third, it assumes a separation of production and consumption, such that nature produces and humans consume. In absolute contrast to these assumptions, my teachers worked with an assumption that they are part of the world around them. As many people said, they were born from and for their Country. Further, they understand themselves to be participants in life processes, and understand life processes to involve other living things as well as the broader seasons and elements discussed earlier in this chapter. They understand themselves to be the beneficiaries of many parts of the world around them. At the same time, they understand other living things also to benefit, and they understand benefits to be mutual and reciprocal. Much of this understanding can be summarised in the context of Country and expressed in the foundational moral principle; a Country and its people take care of their own (see also Rose 1992, 106–10).

Jessie took me walkabout. Sometimes I walked in her footsteps, sometimes we went hand in hand. Her history was laid out on the Earth around her in the sites and stories, in the land and in her memories, and in the unpredictable events of the living world. As we travelled, we saw all around us the roads, fences and 'no trespassing' signs, the erosion, the scald areas, washaways and caved-in river banks that told of conquest in action. I have met and worked with many people, settler descendants and Indigenous, who are concerned about what is happening to the place, but I know of nobody who takes care of Country with the holistic passion that senior Aboriginal people like Jessie bring to their lifelong practices of care.

This book is an exploration of walkabout. It has as its core the ethnobotanical and related work that I have done, and is built up out of experience, knowledge and memory. Memory is important because both the Country and people's lives are changing rapidly. While ethnographic from start to finish, this book also contains much environmental history. It tells of things that settlers cannot know without the help of Aboriginal people: what used to be here, where it was, what was its beneficence, its Dreaming, its sites of increase, its connectivities. This is a book of survival and loss, of love, care and accommodation. And so, it is also a study of Indigenous philosophy examined through an ecological lens. Ecology is the study of connectivities in the living world; this book follows issues of connectivity into domains, place, time, ethics and nomadics.

Nomadics

Aboriginal Australians were for some 60,000 years or more nomadic hunter-gatherers. Their root paradigm for sustaining and understanding life worked with mobility. Their territoriality articulated both their belonging and their mobility. Aboriginal people thus engage with a particular nomadic problematic: that of being here and not-here at the same time, of being both localised and mobile.

'Place is pause', according to Yi-Fu Tuan (1977, quoted in Flores 1998, 31). He means to point to the idea that place comes into being because human beings stay there long enough to articulate meanings for it. Without denying the importance of this emphasis on pause, it is significant that Victoria River Aboriginal people would not define place in this way, nor would they adhere to Tuan's other great dictum, that place is 'space plus people'. This dictum, similar to the former, asserts that humans make the meanings that distil place from pre-existing space.

In contrast, my Aboriginal teachers speak about and act toward place in a dialogical mode, so that place can be understood as a site of multiple presences and encounters. The sacred geography of Aboriginal Australia is wonderfully inclusive: the Country, the waterholes, the increase sites for plants, animals and humans, the fishing places and yam grounds, the birth and burial sites, the Dreaming sites and tracks, the histories, the stories, the songs, dances and much more are part of the geography of Country. The lived time of ephemeral life unfolds in real and located (not geometric or imagined) places.

Place is also intensely political. There are people for Country, and Dreamings for Country, generations of deceased people and unborn people for Country; there are other species who belong there, and there is vast knowledge (daily as well as ritual, secret as well as open, formal and informal) that belongs to the Country. The people and other living things who belong to the Country are usually spoken of as 'owners' of the Country and of the knowledge that is crucial to the sustenance of the Country, the sites and tracks, and contemporary Dreaming action in the world.

I have discussed these and other matters at length elsewhere (Rose 1992, 1996). The issues I raise here are widely discussed both in the literature and because they form the basis to almost every Aboriginal claim to land heard under the *Aboriginal Land Rights Act (Northern Territory) 1976*, in court records. The right to sing the songs of the Country, to lead the rituals of the Country, to dance the Dreamings of the Country through, or bring them up from within, the Country are prerogatives of those people who are identified as 'owners' under the system of ownership that obtains in a given area. Ownership is actualised in the exercise of responsibilities; it is best to think of rights primarily as rights to responsibilities. That is, to be an owner is to hold certain duties of care, to have the right to exercise those duties, to collaborate with others in the exercise of duties, and to prevent misuse by people who are not entitled. Duties of care arise out of relationships. Rights, in this most fundamental sense, are rights to be in and to sustain relationships.

There is an ongoing debate about whether the term hunter-gatherer (or some variant thereof) actually refers to a category at all, and if so, whether the term is appropriate. Annette Hamilton (1982) takes up this issue, noting that it really is a residual category. That is, it includes all the peoples who are not cultivators, wage labourers and so on. Lourandos (1997) also offers a critique which requires him to transform categories into a continuum. He is then able to show overlapping areas along a continuum between hunter-gatherers of Australia and hunter-horticulturalists of Papua New Guinea. The boundary between hunter-gatherers and their nearest neighbours in terms of subsistence strategies is not clear-cut but rather constitutes a cline.

A further aspect of these debates concerns the quality of relations between hunter-gatherer peoples and their environments. Nurit Bird-David (1990) has been a major contributor at the conceptual level, and in a series of articles has argued for a third kind of economy: in addition to commodity economies and reciprocity economies, she suggests a giving economy. Based on her work with hunter-gatherer peoples of India, as well as her reading of the literature, she suggests that many hunter-gatherers treat the environment as a friend or relative: a sharing partner with whom relations are founded in trust (1992a, 31). In another article she proposes that the activity is not well termed hunting-gathering, that it's a mistake to try to force it into a production category, and that it would be better termed 'procurement' (1992b). Ingold (1994) endorses Bird-David's (1992b, 40) case for the appropriateness of the term procurement in its connotative range of 'management, contrivance, acquisition, getting, gaining'. In yet another paper, Bird-David (1993) looks comparatively at metaphors of relations between peoples and their environment. Here she looks at forms of relatedness, or metaphors of relatedness, and considers Warlpiri people to be related to their Country through the procreative powers of the ancestors.

While I think this is too simple, and the article is full of inconsistencies, Bird-David nevertheless offers an insightful reading of the available evidence;[2] I take these issues further in broadening the analysis to concern the proactive dimension of human and other living beings (discussed further below) so that procreation is engaged in across species and domains of Dreaming and ordinary life.

Neither 'production' nor 'procurement' does justice to Aboriginal Australian ecological interactions. There are two aspects to the issue: the bringing forth of life and the sustenance of life. Bringing forth would seem to be locally prior, since without life there would be no sustenance. On the other hand, without sustenance, there would be no life to bring forth life. The relationships between bringing forth and sustaining are interactive and mutually generative.

Ingold's finest contribution in these debates is his essay *From Trust to Domination*. In this work, Ingold (1994) offers a critique of the western paradigm of production in which nature provides the substance and human reason provides the form, so that production lies in the inscription of form upon substance (5). Hunter-gatherer peoples do not inscribe form, nor do they 'exploit' resources; rather they keep up a dialogue with the environment (11). The relationship in which this dialogue takes place is one of trust. Trust, he says, involves dependence without loss of autonomy (14). Trust is not a metaphor; it is literally the quality of relations that hunter-gatherers have with constituents of their environment. Ingold asserts that they do not separate out human from non-human agencies: there is just one world, embracing humans, animals and plants, and landscapes (Ingold, comment in Bird-David 1992b, 42).

Ingold asserts that 'the real challenge of hunter-gatherer studies is to develop conceptual vocabulary that will enable us to capture the dynamic potential of a radically alternative mode of relatedness' (comment in Solway and Lee 1990, 131; see also Ingold 1996). The difficulties go beyond conceptual vocabulary, and the metaphor of capture may be inappropriate, but I agree completely with his view that in our encounters with hunter-gatherers, we encounter a radically alternative mode of relatedness between people and the world. I seek to understand this relatedness through the root paradigm of nomadics.

---

2   Debbie intended to discuss these inconsistencies here but did not do so—eds.

~ ~ ~

In recent years a public misconception has arisen concerning nomadism. The idea seems to be that nomads are free to travel here and there as they see fit, and thus that their purposeful activity can actually be described with the vague term 'wandering'. This view completely misrepresents nomadism as a form of subsistence activity and not as a serious way of life. It conflates nomadism with transience (Flores 1998, 31), perhaps in the nostalgic fantasy that contemporary diasporic peoples, global tourists and corporate transients can be sensibly equated with nomads (see also Rose 2000).

Aboriginal nomadics, unlike transients, are set within the political, ecological and sacred geography of Country. Place is not pause, as Tuan (1977; quoted in Flores 1998, 31) would have it, so much as it is return. Living things are coming and going, and, most importantly, returning. The Aboriginal problematic is beautifully stated by Stephen Muecke in the dedication of his co-authored book *Reading the Country:* 'To the nomads of Broome, always there and always on the move' (Benterrak et al. 1984, 3). There is both movement away, and movement toward, both the departure and the return. A Country includes a plurality of places, and people have rights to travel in, learn about and undertake ceremonial and daily responsibilities for more than one Country, so that the 'range' (to use Stanner's 1965 term) is quite broad. But at the same time, it is bounded not only by social and ecological limits, but also by an ethic of return that keeps people 'always there' as well as 'always on the move'.

Aboriginal nomadics work with the departure and the return, and this same ethic underpins much of the communicative quality of Country, as I discuss in Chapter 6. The point to bear in mind is that Aboriginal mobility is not a matter of being pushed from one depleted resource to another in an endless quest for survival. Rather, people conceptualise their mobility in a communicative mode. The Country tells what is going on, and people respond to messages of Country. Rather than being pushed by necessity, they are called into action by communication.

Concepts of presence and absence take on a specific gravity in this culture of motion and return. Presence is an embodied emplacement and, as I will show in Chapter 2, is relationship. One is always present to and for others. Absence is the state of being somewhere else. It is not an ontological empty space, but a departure set within an ethic of return.

2

Footwalk Epistemology

Two different law. [One for whitefellas, and] Nother
Law belong to *ngumpin,* blackfella. You know him
bin walking on this land for many many years.
Him bin walking by foot.

                                      Riley Young Winpilin

Aboriginal hunter-gatherers make efficient and effective use of their
Country, and effective use of course dictates movement across the Country.
Such movement is also embedded in an ethical context. The Country is full
of messages that call to people. Victoria River nomadism was not usually
a matter of being pushed from place to place as resources become scarce,
but rather of being called from place to place as food comes into plenty.
In land claims across the Northern Territory, from desert to islands, people
speak of nomadic travels as joyful engagements with plenty. Victoria River
people, like others, have stories of the hardships of drought and flood, and
there are sites of refuge as well as stories of people who did not survive, but
people speak of these events as exceptions. Mostly they speak and sing of
sounds, smells and colours, of ephemeral actions such as march flies biting
or plants flowering, of the plants and animals that form this communicative
matrix. Sensuous information is a call to action, beckoning those who
know what is happening. Country comes into being through the calls and
responses, the departures and returns of life and life-giving action. Both a
communicative matrix and a matrix of action, it lives in time and unfolds
into relationships that promote serious life.

By foot

In North Australian Kriol and in Aboriginal Pastoral English as they are spoken in the Victoria River District, to go walkabout is to go out getting food. It might be fishing, hunting, digging, picking, cutting or chopping, and most probably would be some combination of techniques, employed opportunistically as events unfolded. You might take a spear, a rifle, a digging stick, your fishing gear and a hand axe, known locally as a tommyhawk. Women rarely use spears or rifles; the digging stick ('crowbar') is their iconic tool. Fishing gear and tommyhawks are used by everyone. Jessie had a carrying bag which held her essentials, and she often carried her little tommyhawk if there was any likelihood of finding native honey (sugarbag).

Walkabout is never aimless. It depends on knowledge, rights and responsibilities—one goes walkabout in Country one knows and has rights to be in; one goes with intent, and one announces that intent. The term walkabout has entered Anglo-Australian English in almost the exact opposite of its Indigenous meanings. Whitefella Australians used the term to refer to Aborigines who abandoned their jobs in order to return to Country, thus implying an abdication of responsibility (Arthur 1996, 173–74). They also used the term as a descriptor of aimless movement: 'walkabout disease' is a form of poisoning that afflicts horses who eat one of several toxic plants and then become disoriented and wander relentlessly (Toop 1958). Applied to Aboriginal people, the term suggests an uncontrollable and aimless urge for motion. On the other hand, in the pastoral world prior to 1967 or so, when Aboriginal people were turned off the stations in the wet season, they were sent walkabout, as discussed briefly in Chapter 1. This usage did not necessarily imply aimlessness, but it served to erect a conceptual barrier between Aboriginal action in Country and 'white' productive activity. In English spoken by Europeans on pastoral stations, walkabout was a holiday, and thus was clearly distinguished from the pastoralism that was held to be serious productive work. Aboriginal people, too, sometimes refer to walkabout as a holiday, clearly distinguishing the autonomy of life and knowledge in the bush with the servitude of station life.

In Anglo-Australian English the term walkabout strongly implies actual walking, but in Victoria River Aboriginal English the term means leaving camp and going for food. It always implies the return, it always implies productivity, it always implies purpose. It need not, however, imply walking.

You can go walkabout in a truck just as well as on foot, and in the past two decades as motor vehicles have become more equitably available, people often do go walkabout by truck, at least for part of the way.

When you go on foot, you 'footwalk'. In Aboriginal English the term walk includes many forms of locomotion; footwalk speaks precisely to the feet on the ground. Across the north, Aboriginal people hold that footwalk offers a privileged form of knowing. An Aboriginal man in the Kimberley explained to Richard Davis (2005, 159) that he and his countrymen 'know the land because they have walked it, not like consultants, they don't know the land, the way the water flows underground'. Footwalk connects body and Earth; the feet mark the Earth and those marks become traces of one's presence. In the days when the foot was the only form of terrestrial locomotion, all knowledge was footwalk knowledge (using the term generously to include swimming across rivers and other forms of unmediated motion). Today it is always essential to approach sacred and dangerous places on foot, and footwalking in Country is regarded as a key element in people's education.

Chris Healy writes about walking as a form of knowing along the Aboriginal tourism and Dreaming walking track in Western Australia, the Lurujarri Trail. In reflecting on experience and refusing theoretical closure he both performs and articulates an

> ethic that privileges getting to know the land simply by looking and seeing and feeling the place rather than by 'showing', reading, Socratic dialogue or the routines of investigation that are the staple of intellectual, scientific and juridical inquiry. (Healy 1999, 65)

I take a similar path in exploring a footwalk epistemology. My intention goes to a set of questions that concern the embodied motion of walkabout. Motion directed toward a purpose is intentional action; I am asking about purposeful movement in Country. What is the shape of knowledge gained in motion? What are ethics in a world of motion? How does time work in a world of motion? What is motion when it is directed toward Country? What kinds of ecological understandings emerge from motion? Is motion a generative force, and if so, how and why?

Answers to these questions will emerge in the chapters to follow, and later I will expand my questions to include ideas of footwalk ethics and footwalk creation. For now, I will stay with epistemology. A footwalk epistemology is embodied: you are on the ground in your person. If you are barefoot, you feel the texture and temperature beneath your feet but, in any case, footwalk epistemology requires your bodily presence. As you are there in your body, footwalk epistemology engages your full sensorium. You are, after all, seeing and breathing, your skin is enveloped by air, heat, humidity, wind or breeze. Ants and mosquitoes walk on you, and march flies bite you. As you walk, prickly trees scratch you, and branches slap. The prickly spinifex can draw blood from your ankles, and stones and vines can trip you. You watch, listen, talk, sniff and wonder. You are there as a thinking, experiencing creature, and so you take notice. At the same time, you are in the presence of other thinking, experiencing creatures. You are noticed.

Another aspect of footwalk epistemology: you are in motion. You are going somewhere, and you have come from somewhere. You are emplaced. Footwalk knowledge may be specific, or it may be generalised, but it is always grounded. Another aspect: you are in time. Your motion is taking place in time and as you come from somewhere, so you have a history, a memory, a fund of knowledge that relates to where you are *not,* as well as to where you are. Then, too, you are going somewhere. Along with memory you bring your expectations, your foreknowledge and your purpose. Emplacement situates you; time connects you to many places.

Footwalk epistemology is never isolated. Embodied, emplaced, situated in time, you are also in the presence of other living things. Some are human, some are extra-human, there are ancestors and Dreamings, and there are all the ordinary living things—the trees, birds, insects, marsupials and mammals, fish, amphibians, reptiles, trees, shrubs, grasses. To some of these you are related. So, you are in the presence of your relations, and thus you are enmeshed in reciprocities, obligations, benefits and shared objectives. You are not alone.

You are not alone, and thus you are enjoined to ethics. Not only the morality of observance of Law, but a broader ethic of connection, belonging and encounter, call to you in this way. Footwalk epistemology arises in the in-between. It is motion connecting the here and the not-here. On walkabout you are between here and there, you are in motion rather than stasis, you are in a state of encounter, and thus in flowing moments of presence. On walkabout you are enmeshed in connectivities: between self and the other selves of the encounters; between self and place; between memory and event. If your hunting is successful, you are also working between life and death.

The motion and the embodied presence give rise to authoritative knowledge. The knowledge of old people, for example, is said to arise not only from time but also from place: they know a lot because they have 'been here' for many, many years. Their knowledge is thus a part of the here and the not-here, the motion and the return. Big Mick Kangkinang was a man of impressive knowledge; Darrell and I privately referred to him as the old man who knows everything. He explained that the knowledge of old people who have travelled widely is a map. To have a big map, or the biggest map, and to have the right map for the place where one is, is to hold invaluable knowledge. He explained that when he was a young man his father took him all over the place. That's why he's got 'the right one map'.

The map tells of origins and Law. Jimmy Manngaiyarri explained:

> I mean, you know, on this Earth. On this ground ... You know, every Aboriginal people know, what the Dreaming bin do. This ground, he's sort of a map. Map. Map for the people. That way, this one ground. We got to follow that. Don't matter where we are, we got to follow. By Dream. What the Dreaming bin do, we got to follow.[1]

Riley Young Winpilin framed the political message of located knowledge, challenging 'government' to pay attention to Aboriginal people because of their specific epistemologies:

> You know why blackfellow know this land? Him been here many many times over this Country. He walk it, he riding horse, and him been muster, he know where the creek, he know where to go, he know where everything. Why government can't realise this?[2]

The locatedness of knowledge is also signalled by the concept of 'sitting', which means staying in place. My teacher Hobbles Danaiyarri told long stories of colonisation that were addressed to white people (see Rose 1984), and in one of his narratives he implicitly spoke to the widespread perception that settler Australians are confused when confronted with contentious issues. He told them: 'If you want to believe, believe for the Aboriginal people. He's the one sitting on the land.'[3]

~ ~ ~

---

1   Jimmy Manngaiyarri, tape 110, recorded at Yarralin, 13–14 August 1991.
2   Hobbles Danaiyarri, tape 16, recorded at Lingara, 11 April 1982.
3   Hobbles Danaiyarri, tape 1, recorded at Yarralin, October 1980.

All this emplaced and time-binding motion emerges from a cosmogony of movement. Creation starts with the travels and actions of the great creative ancestral, or Dreaming, beings who walked the land and sea. All through their travels the Dreamings brought into being the differences that matter. The Australian continent is crisscrossed with the tracks of the Dreamings: walking, slithering, crawling, flying, swimming, chasing, hunting, weeping, dying, giving birth. They were performing rituals, distributing the plants and marking the zones of animal distributions, making the landforms and water, and making the relationships between one place and another, one species and another. They were leaving parts or essences of themselves; they would look back in sorrow; and then continue travelling, changing languages, changing songs, changing identity. They were changing shape from animal to human and back to animal again, and they were becoming ancestral to particular animals and particular humans (totemic groups). Through their creative actions they demarcated a world of difference, and they made the patterns and connections that crosscut difference.

Dreamings travelled and they stopped. They stopped, they changed over into permanent sites and into other living things, and they stayed. In stopping and changing over they became ancestral: Owlet Nightjar Dreaming, for example, became the owlet nightjars and the owlet nightjar people of today. Equally, however, they kept going. Dreamings are masters of the art of being here and not here at the same time. They also are both then and now: both origins and contemporary presence. People interact with Dreamings in daily life as they do their hunting, fishing, gathering, visiting and resting. And people interact with them powerfully in ceremonial contexts which I will discuss later.

Footwalking with Ivy

Ivy Kulngarri and I worked and travelled together in the mid-1980s to document sites and other matters pertinent to Indigenous ownership of Country. Her home settlement in those days was Pigeon Hole; her language was Bilinara, and her Country was crisscrossed by several great Dreamings, including the creator women known as the Nanganarri Women. These great Dreaming Women brought the people, Law and ritual for this Country; and their presence empowers women today. Stories, designs and songs tell of Dreaming actions, as does the land itself. Where they travelled, where they stopped, where they lived the events of their lives, all these places are sources and sites of a dynamic gendered system that is called Law in Aboriginal English.

Figure 2.1. Ivy Kulngarri, Nancy Jalayingali and Mollie at the bullwaddy tree at Kamanji, Pigeon Hole area, 1984.

Source: Photograph by Darrell Lewis.

We made a trip to a billabong where the Nanganarri Women had stopped and bathed. It is a beautiful little pool of water tucked away in an otherwise dry and scrubby area. Pandanus trees surround the billabong and are themselves Nanganarri Women transformed; their roots and other aspects of the billabong are deictic features of women's bodies. Ivy and I saw the remains of an old campfire. Lying on the ground amongst the charred sticks were the rusting remains of some burnt tin cans. I felt shocked: rubbish, litter and junk right here at the place where the Nanganarri Women walked and where they are, even now, today.

We were standing near a billabong that is sacred to the Dreaming Women. Ivy looked at the tin cans and she burst into a happy smile. She pointed to the remains of the fire, and she explained:

> That fire, we had a dinner camp, oh, long time now, we came walkabout, and we got fish, and we had a dinner camp. You remember Roy, my son Roy? Roy was here that day. We had a dinner camp here, we got fish, and we had damper and tea, and fish. Only Roy, Roy had tin of beans. That tin there now, Roy had that one, baked beans. We had dinner camp right

> here, and Roy had that baked beans. After, when we
> went back, they grabbed Roy, and took him away
> for young men's business. You remember that ring
> (ceremony) place, eh? You were here then.[4]

This is a Dreaming place, and it is Ivy's home. She lives in her Country, raises her family here, does her Law business here. The remains of the fire tell an intimate story about Ivy and her son Roy, and their fishing trip, so it is a story of kinship and nurturance. Ivy nurtured her children, and their Country nurtured them all. They got fish. They cooked and ate them, and they left the remains. Dingoes might have come round later, and crows, for sure. No doubt the ants worked there too.

This is also a story of belonging; they were here because they have responsibilities here. This is where they belong. The place holds the traces of the lives that have been lived here, and the traces tell the stories, but the stories are not unambiguous. The story of belonging can only be apprehended in the first instance by those whose presence here is called forth by their relationships with the place itself.

My initial reaction emerged from a western story of keeping places beautiful by keeping them tidy. According to government publications, the concept of litter, defined as waste improperly discarded (in public places), is a fairly recent invention. In Australia it is treated first and foremost as a problem that combines aesthetics with citizenship: 'Litter is unsightly and is visible evidence of antisocial behaviour' (Australian Environment Council 1982, vii). The idea is that a good citizen picks up after herself; she keeps Australia beautiful by erasing the traces of her presence. Her citizenship and her care of place are demonstrated most responsibly by her absence.

I hold no brief for litter; it is a serious problem in both Aboriginal and settler communities, and it impacts on human health and the wellbeing of other species. However, the idea of self-erasure is far more problematic. Ivy absolutely does not erase herself. Rather, she announces herself. When they went to that billabong that day, she would have called out to the Dreamings to say who she was, and she would have called her ancestors saying: 'Here we are, I've brought the kids, we're hungry.' She would have used the kin terms that connect her to her deceased relations. The Nanganarri Women made the Bilinara language and left in the Country the future generations of Bilinara people. Ivy and her children originated from Nanganarri action

---

4   Taken from rough notes—eds.

and are known by the Dreamings. And so, by human ancestors and by Dreamings, the Country would know them, and would feed them. We know that they were cared for: the remains of the fire are the proof of her Country's nurturance.

Ivy made this fire here quite a few years ago. She knew, but probably Roy did not, that this was Roy's last day as a child, so this is also a story about gender and Law. The story reaches out through space to neighbouring communities and neighbouring Countries. Many people came to Pigeon Hole to work to make Roy into a young man. Perhaps Ivy had taken him walkabout to get him away from the community; perhaps she wanted a last day in the bush with her little boy. When they returned to camp, Roy was grabbed.

This story leads back to the billabong and the tin cans, and it reaches out across people, places and times; it reached out and captured me as well. A few days after they took Roy, they brought him up to Victoria River Downs (VRD), and all of the mob I was with at Yarralin went over for the business. Along with a lot of other women, I worked all night to effect Roy's transformation from little boy to young man. The men had made a ceremony ground, or ring place, by inscribing concentric circles into the ground to mark the area for singing and dancing. The men sang and we danced, and the marks of this event, the tracks of our feet cutting across the inscribed circles, remained until the rains washed them away.

Places are full of presence, both present presence and absent presence. Charred sticks and tin cans hold a story that extends through time, but the extension is not infinite. Ivy and Roy, and that fishing trip, are present in this place for as long as the traces of the day remain. Ivy occupies her Country by inscribing in it the passage of her life. The places hold her presence in the periods of her absence. Living things return, and leave again, crossing time and space. To live morally one must return and return again. Traces of presence are thus knots in the dynamic of interlocking places. Such traces are time binders: they mark the returns as separate events and thus articulate time as difference. Similarly, they bind time and place into the lives of living things.

My Aboriginal teachers speak about and act toward place in a dialogical mode, so that place can be understood as a site of multiple presences and encounters. Place is the product of the lives of many living things, including extraordinary beings and non-human beings. Thus, the lives of animals and plants, of insects and rainbows all contribute, in their coming and going, their presence and non-presence, to the life of the place. Footwalk epistemology works in the encounters between real people with their multiplicity of connections and other parties to those encounters. Place, I am saying, bears and holds the marks of encounter, and thus becomes densely saturated with intersubjective presences.

Presence

Thus far I have used the words 'presence' and 'absence' quite casually. I now want to bring contemporary western philosophy and theology into conversation with the kind of presence entailed by walkabout. Daniel Boyarin, in his magnificent essay on ocular desire (1990), takes issue with the idea that Judaism introduces to the world the idea and ideal of an unseen God. In contrast, he points to the many places in Torah where God is seen, particularly in the book of Exodus, where Moses encounters God face to face, and others see at least some part of Him. This face-to-face encounter with God is, in Boyarin's view, an encounter with ultimate Presence. Hereafter, says Boyarin, those who love God long for His Presence, and seek to overcome his absence through a recuperative hermeneutic, leading into a moment when God was not absent.

Boyarin differs from many philosophers and theologians in treating God as an embodied being (thus seeable and knowable), but his work points to a central nexus of thought in the west: loss, absence, separation. One strand of Christianity can be seen to follow this path: God walked in the Garden of Eden with Adam and Eve, but when they were expelled, they were banished not only from the Garden but also from the daily presence of God. Christians in some periods saw nature as God's creation (e.g. Gurevitch 1985, 57), while another strand of Christianity devalued the world because it was 'fallen' (Merchant 1980). According to some views, God's hiddenness was overcome in Christianity in the person of Jesus; and so, Michel de Certeau (1992, 81) argues, the missing body of Jesus haunts western Christian imagination. The foundational moment of Christianity is the disappearance of the body. Marcel Gauchet contends, not dissimilarly perhaps, that Christianity signals the exit of God from the world. His point is that Christianity imagined God as so beyond human experience that He took a human form in order to have contact with humans at all. Nothing, it seems, that humans could experience on Earth would offer a trace of God (Gauchet 1997).

Plato posited another form of loss or separation. This is the separation of spirit from body, and the organisation of value such that full value resides in a transcendent plenum of spirit, while the remaining material world is of lesser value (Plumwood 1990, 1993). According to Stephen Tyler's (1984, 34) analysis of the dominance of vision in western thought, Plato's ontological gap between the real world and the ideal world created an inconsolable sense of separation from the real. Plato, along with other founding figures in western thought such as St Paul, works with modes of thought that are pervaded with loss, and their longing is to attain presence—to close the ontological chasm and be returned into the one. Susan Handelman describes Paul's project vividly: 'To rend the veil, attain the pure presence of the ultimate referent, collapse differentiation, bridge the gap between all signs and the ultimate signified' (Handelman 1982, 89).

Against this sense of loss and longing for return into undifferentiation, the Enlightenment appeared to offer another path entirely: that of the ultimate knowability of the universe, and the ultimate achievement of human satisfaction. Zachary Braiterman (1998, 143–44), for example, asserts that the term 'presence' has been synonymous with 'logos' in western thought and that it thus refers to certainty, communicability, identity, encounter with God, end to alienation and redemption. Postmodern theologians advance counter-themes of absence, fragment, deferral, dispersal and difference. In a post-Holocaust world marked by rupture, violence, eradications and extinctions, the concept of a transcendent plenum no longer carries conviction. If God registers at all, it is as a partial and fleeting trace (Levinas 1994).

A view of presence that locates the greatest Presence of all in a transcendent plenum suggests a definition of presence that is as far removed as I can imagine from the presence I refer to in an Aboriginal context. I do not intend the term to mean *either* distant and desired transcendence *or* the broken and fleeting fragments of a lost transcendence. I use the term rather to speak to the particularities of living things in their intersubjectivities. Presence is thus always located in the world in living beings and in places. Presence is not all-encompassing. One is present to others, not subsumed within a greater presence. Therefore, to be present is always to be in relationship: one is always present to another. There is no desire for undifferentiated unity; to be present is always to be particular. Particularity involves embodiment, emplacement, history and ethics.

My use of the term 'absence' follows from my definition of presence. Absence denotes the state of being departed—it is an absence known by contrast with presence (see also Fuery 1995, 1). It is not to be construed as an existential or ontological empty space, but as the consequence of mobility. Absence is the not-here that alternates with the here. In this usage, absence is part of an ethic of multiplicity and motion. To return one must depart, and thus ethics in motion entails absence as an alternative to presence.

Perhaps I also need to clarify my use of the term 'world'. In some western traditions, the world has been classed as a poor relation to the transcendent plenum known as God. Whether the world is fallen, or abandoned, or simply the lesser reflection of eternity, the term can invoke connotations of lesser value: the body as a lesser thing than the mind, the material as lesser than the spiritual, the world as lesser than God. It will become clear that I am not dealing in these binaries at all. I aim to engage with an ecology of place that works with living beings and places who are present to each other in the particularities of their being and in the intersubjectivity of their becoming.

The implications of this difference really matter. From a postmodern theological position, Richard Rubenstein argues that humanity had best forsake the quest for redemption and accept the 'what-is' of this world (in Braiterman 1998, 99). His position raises significant questions concerning what is the 'what-is' of the world, but my more immediate point is that he seems to offer engagement with the world as a default position to which we are thrown back after experiencing the absence, loss or hiddenness of God. I raise this point because I think it is profoundly western and yet often quite obscured. The foundational event of Jewish and Christian culture is expulsion from Eden. For the descendants of Adam and Eve, the world always already is a second-best option, the position that is there when your preferences disappear. Carolyn Merchant (1980) traces the expulsion story into today's shopping malls via a route that follows science in claiming the ability to know and to remake the natural world. That is, she sees the contemporary world struggling in the wake of an Enlightenment paradigm that asserted humans' ability to remake the world into a better and more manageable place. The 'what-is' of this Enlightenment paradigm is a set of processes to be discovered and problems to be overcome. In the wake of the Enlightenment, in the post-Holocaust west, the 'what-is' of the world is a broken and wounded terrain of catastrophe and risk, as well as being, of course, a world in which modernity continues to wield great power with many mixed benefits.

It could be said that my Aboriginal teachers also engage with the 'what-is' of the world. Their 'what-is' is an ecology of crosscutting and overlapping connections that call living beings into emplaced intersubjective becoming. The care of the world *as it is* is not a default position undertaken in the absence of God or in despair over the world, or out of dissatisfaction with the given, but rather is the vibrant practice of engaging one's living presence with the world of living things. This is not to say that they do not recognise catastrophe and damage; rather the 'what-is' of the world contains rain, drought and floods, plenty, scarcity and serious hunger, life, death and continuity.

The 'what-is' of this world is creation, Law and all the conditions that enable serious life. The 'what-is' comes from the Earth itself. One side of Dreaming is that which creates and endures, and the other side is this ephemeral world: the living things, the relationships between and among them, the waters that support their lives, the cultural forms of action and knowledge that sustain the created world. Aboriginal people's daily lives, as well as their ritual and other forms of care, unfold in an ecological poetics of connection. The work of creation continues to happen in the world precisely through the ephemeral. Both daily and ritual work seek to ensure the continuous flourishing of ephemeral life. Dreaming is thus actualised in present time; the perduring life of creation is carried in contemporary time and place by ephemeral life forms.

Law is Dreaming, and it is literally the Law of land and water, of all life, of connection and relatedness. Law is about how the world works; the term is *yumi*. '*Yumi,* that's Law for Dreaming side all the way—what Dreaming bin done before, you can't lose that Law. Our Dreaming bin do that, we got to hang on [do the same].'[5]

Dreamings imprinted themselves into the world, and thus the world becomes the source of knowledge. According to Jimmy Manngayarri:

> What he did, world bin made, well he got to stay like that. He can't broke that Law. No! That's the way the *ngumpin* [Aboriginal, in this region] Law. You know. Just like, you know, *kartiya* [non-Indigenous people], he go different his school, eh? Well same as the *ngumpin* on this Earth. This Earth give him all the idea.[6]

---

5   Doug Campbell (source not recorded—eds).
6   Jimmy Manngaiyarri, tape 110, recorded at Yarralin, 13–14 August 1991.

Connectivities

The 'what-is' of the world is held together by connectivities. Kinship is a primary mode for talking about relationship and, like other patterns of differentiation and connection, kinship both separates and connects. Plants provide an important and unexpected context for exploring connectivities.

The vast majority of plants that my teachers and I looked at in the course of our ethnobotanical surveys are named separately, just as the western Linnaean system identifies separate species. My experience is similar to that of others in Australia and elsewhere: at certain levels there is a significant correlation between 'scientific' and 'folk' taxonomies. This is an extremely powerful fact. It indicates that cross-culturally humans are attentive to many of the same cues in their efforts to discern differences, to categorise and to classify. That we attend to many of the same cues suggests that there is a fundamentally detailed and accurate quality to classification that is built upon overlapping ranges of observations and interactions. This fact tells us something important about humans: about how our observations and interactions with the world are fundamentally similar at certain necessary and important levels. It also tells us that the idea that the world is only knowable through language must be inadequate. If knowledge was all cultural, if the ability to know was based solely on language, then there would be no reasonable explanation for the extraordinary convergence of systems of classification cross-culturally. The fact of convergence also suggests that there are regularities in the world that impress themselves on people.

Attempts by European peoples to make sense of the Australian flora offer a test case. By the eighteenth century, botany had become an important aspect of colonisation. Captain Cook had with him Joseph Banks, a noted naturalist specialising in botany. Augustus Gregory, who explored the Victoria River District in 1855–56, was accompanied by the botanist Ferdinand von Mueller. From my convergence perspective, I am struck by the fact that right from the start botanists were able to work with a system developed in Europe and begin to make sensible classifications of Australian flora. I am not overlooking the fact that the early classifications required a great deal of subsequent refinement, and that there is always more work be done, and that some aspects of Australian ecosystems posed very interesting difficulties (Robin 2005). My point is that the world is not infinitely variable. There are patterns to the forms that life takes, and those patterns are widespread. Further, humans are part of those patterns: our ability to recognise and make sense of the patterns we encounter indicates that pattern recognition is constitutive of our human capabilities, and thus of our engagements with the world.

A good scientific account of a species includes a broad classification (tree, shrub, grass, etc.), along with descriptions of the bark, the leaves and flowers, the growth pattern, the habitat and flowering patterns. These same indicators inform much of the Indigenous system of classification. To make a set of classifications is not the same as to make a set of meanings. The convergence of classifications between colonisers and colonised was by no means a convergence of meanings. Gregory saw rolling downs and grassy plains and wrote to extol the magnificent grazing lands of the Victoria River District. Aboriginal people did not see grazing lands, although they kept the savannas open for herbivores. Their landscapes were mosaics of land systems and plant communities within which a multitude of plants, animals and human groups thrived. These landscapes were made up of eco-places with their own histories and imperatives.

The two systems of classification diverge in their meanings, and they are not fully identical even in their classifications. The Indigenous system makes a few distinctions at the level of terminological classification that are not made in the Linnaean system. An example is *Terminalia platyphylla*. My understanding is that this one species (Linnaean system) is terminologically distinguished in the Indigenous system. *Marntayark* (*T. platyphylla*) grows along the rivers; it is named for its edible gum (*marnta*) and also provides food for black cockatoos. *Pijpangu* (which also seems to be *T. platyphylla*) grows around springs. It has the same uses. Ivy Kulngarri explained that it has the same leaf and same 'everything', but it is for spring Country.

More commonly, a single species is undifferentiated by name but can be differentiated by habitat or other factors if required. For example, the aromatic *manyanyi* (*Streptoglossa odora*) grows on the black soil and in the stone Country. The black soil type is the main or unmarked type, and the stone Country *manyanyi* can be specified as *wumerangarna* (stone dweller). Similarly, a Melaleuca species, *kungun* (undescribed in the VRD), is differentiated as large and small, the small one being for stone Country.

Unlike the Linnaean system, the Indigenous system does not aim for exhaustive separations. It classes together numerous species that the Linnaean system differentiates. This is particularly evident among the grasses and forbs. Those which have both distinctive features and a known use are named separately. The remainder are commonly classed under the generic term *yuka* (grass). The category *walayinkarri* is a real catch-all: several species of vines and trees are all classed as *walayinkarri*. Other inclusive terms are descriptive; *wuju* (sharp grass) includes several sharp grasses, but not spinifex. *Pakamalij* (having *paka* or prickles) and *murulumpu* (generic 'cheeky') both refer to plants with prickles and thorns, while *pirrpul* labels a number of scrubby acacias.

~ ~ ~

I have examined the major similarities and differences in our two systems of classification because I believe it is important to show that both exist. I have remarked on the massive similarities and have commented on some of the differences. I now turn to a much more fundamental difference. The Linnaean system is built upon levels of inclusion, and with the scientific acceptance of concepts arising out of Darwinian evolution, inclusion comes to imply relatedness to a common ancestor. The system is built on cladistics, a method of inclusion on the basis of degrees of separation from a primal source.

Frank Zimmerman (1996) suggests that many folk taxonomies are built on different principles. Rather than focusing on levels of inclusion, they focus on connectivities. The Indigenous system I am describing here is indeed a connectivity system. Of the 165 plants of the riverine zone which I discussed in detail with my main teachers, fully one quarter are connected to others through a system of relatedness that draws on the language of kinship. Kinship is, of course, the root conceptualisation of connection (Chapter 9). My teachers most frequently used the term 'mate' in talking about connections between plants. The term is delightfully inclusive in its lack of specificity: cross-cousins are mates, people who share the same physical flesh within the system of matrilineal co-substantialities (discussed below) are mates. More expansively, people or others who share a common interest or have something significant in common can be classed as mates.

The mateship connections link up separate species; some are linked within a genus, and some of them cross genera. Within a given genus there are mates that are connected by their similarity (same genus) and their differences in habitat or other criteria. They are 'same, but different', as people say, and therefore 'mates'. Among the eucalypts there are many intra-genera mateships: *narrka* (*Corymbia polycarpa*) lives by preference in the limestone or in the hill Country. It is mate (cross-cousin) for *jartpuru* (*C. terminalis*) that lives out on the plains. *Kunjal* (*Eucalyptus* sp. aff. *argillacea*) is mate for the coolibah tree, *wulwaji* (*E. microtheca*). It is differentiated from the coolibah not only by being its own self, but also by its behaviour: coolibahs grow around billabongs, while *kunjal* is more widespread: it can 'sit down anywhere'. One of the trees labelled *lunja* (*C. aspera*) is a mate for *kunjal* because it is similar and is also said to be mate for *parayinparayin* (*E. miniata*). This latter species does not grow south of the Stokes Range, a major boundary between ecological zones in the region (see Chapter 4), so the mateship here crosses into another zone and links up similar trees.

Other examples abound. *Nampula* (*Ficus racemosa*) is mate for *tinpali* (*F. leucotricha* var. *leucotricha*). Both are fig trees; one is for river Country, and one is for stony hill Country. *Pirkili* (*Melaleuca nervosa*) is mate for *pakali* (*M. argentea*); one is for dry Country; one grows only along the river. Brachychitons offer another example. One kind, *yingki* (*B. diversifolius*) is 'mate' for another kind, *jaringkal* (probably *B. paradoxus*). The differences between the two are habitat and location: *yingki* is a river species and *jaringkal* does not grow in the riverine area under study. *Jaringkal* itself has another mate: *miyaka* (probably *B. multicaulis*): *jaringkal* is the tall one and *miyaka* is the short one. *Miyaka* also does not grow in the riverine area under study, but as I will discuss (Chapter 4), it is emblematic of Mudbura people and their desert Country.

When we turn to mates that cross genera, we find comparable issues of similarity and difference. Within one genus it seems that mates are the same, but different. Across genera, it seems that mates are 'different, but the same'. *Ngamanpurru,* the conkerberry (*Carissa lanceolate*) is mate for *kumpulyu* (white currant, *Flueggia virosa*). Both are shrubs and carry edible berries, one white, one black. Similarly, *panganpangan* (*Terminalia erythrocarpa*) is mate for *mawunji* (freshwater mangrove, *Barringtonia acutangula*). Both grow along the banks of the rivers and creeks, and both produce food that falls into the water and is eaten by fish and turtles. *Manyingila* (*Excoecaria parvifolia*) is mate for *pirijpirij* (*Terminalia bursarina*); they look almost exactly alike, but one grows in the riverbed, and one grows out on the dry ground. Other mates include different trees with identical-looking bark, and two trees, one acacia and one grevillea, that have long needles. When the wind blows through the needles it makes a similar sound in both trees.

Some, but not all, of the mate relationships discussed above were further specified as cross-cousins. This is an interesting relationship for the same/different contrast. Cross-cousins trace their ancestry back to a shared grandparent, and thus are 'same' in the context of the group of descendants of that person. On the other hand, cross-cousins are the parents of children who are the preferred marriage partners and because marriage must take place outside the group of people classified as close kin, cross-cousins by definition are 'different'. If they were not different it would not be appropriate for their children to marry. Cross-cousins are thus 'same' with reference to their shared ancestor, and 'different' with respect to the next generation.

There are a few other relationships among plants that are designated in terms of specific kinship. Another of the trees labelled *lunja* (snappy gum, *E. brevifolia*) is brother to bullwaddy (*Macropteranthes kekwickii*). Brotherhood here speaks to matters that fall within men's business, but the relationship is not random. *Lunja* and bullwaddy alternate as dominant species in the Mudbura *kaja* Country north of the great grassy plains, and while *lunja* is not restricted in its distribution the way bullwaddy is, a familiarity with Mudbura Country suggests that regular and predictable contiguity also counts as a form of connection. Where one lives, the other does not, but in Mudbura Country they are always side by side.

There are other such relationships: *wanymirra* (*Vigna lanceolata*) is 'mother' to *wayita* (*V. lanceolata* var. *lanceolata*). The issue here is size. *Wanymirra* is much larger than *wayita*: mother and child. An acacia called *minyngatj* (probably *A. megalantha*) is the child of *parrawi*, the fish poison (*A. holosericea*). *Cassia notabilis* is mother for *A. nuperrima*, and both are medicines. Among the lilies there are also mother–child relationships.

In sum, patterns of relatedness work with connectivities of similarity and difference, and thus produce a system that is more in keeping with phenetics than cladistics. The abstract pattern is familiar: differences and similarities crosscut each other to produce mosaics of connectivities.

A short walk with cattle

I do not wish to provide a history of cattle in Australia.[7] I limit myself to an overview of the four-legged soldiers and the radical and (often) irreversible changes they and their human masters wrought in Australian ecosystems. I then turn to a more detailed study of ecological change caused by cattle in the Victoria River region. Here I anticipate that discussion in order to make the point that cattle and white settlers have brought a new kind of absence into the Victoria River Country. This is the absence that does not return: an ontological absence marked by an empty space where something that once existed no longer does so. This absence presages 'double death'.

Australian mammals do not have hooves. The millions of years of mammalian life in this continent never impacted upon the soils the way hoofed animals do. Cattle were brought to Australia on the first fleet; they came from Cape Hope and were mainly Afrikander breeds, but within a few years settlers were importing cattle from all over the world (Rolls 1984, 18–19). One of the great early estates was named 'Cowpastures' because it was discovered by cattle who wandered off when their attendant fell asleep. In Eric Rolls's pithy words: 'It was the most successful exploration in the first twenty-five years' (1984, 12, 18–19). Cattle were in advance of settlement in many areas, and the wild cattle, small and tough, took over degraded areas as well as causing degradation. For example, wild cattle moved into portions of Queensland that had been taken over by the introduced prickly pear; their tongues atrophied to the point where they could ingest the cactus. They never drank water and lived only on prickly pear; their gait changed, and they could outrun horses (Rolls 1984, 22).

Throughout the continent, cattle, horses, sheep, goats and other hoofed mammals brought rapid and often irreversible changes. The immediate impacts were on soils and grasses. Rolls describes the changes vividly:

> [The native grass] roots had run in a spongy soil full of humus. They were accustomed to fire, to drought and flood, to deficiency of nitrogen and phosphorus, to the gentle feeding of sharp-toothed kangaroos at their clamped butts, and the picking of their seeds by parrots and pigeons and rats. They had never had their whole seed

---

7 This chapter is meant to provide a brief overview. There is a quantity of excellent and detailed studies which include Beale and Fray (1990); Bolton (1981); Flannery (1995); Lines (1991); and Marshall (1966).

heads snatched in one mouthful; they had never been trampled by ... hooves; their surface roots had never had to run in hard ground. (Rolls 1981, 28)

Destruction of palatable native grasses enabled more hardy and less palatable grasses to colonise widely, and subsequently settlers introduced English and other grasses that were adapted to trampling (Rolls 1981, 28).

The explorer Thomas Mitchell saw the inland plains and recognised the interactions of Aboriginal people, their cultural fires, the nourishing grasslands and the herds of kangaroos. He wrote in 1848:

> Fire, grass, kangaroos and human inhabitants, seem all dependent on each other for existence in Australia, for any one of these being wanting, the others could no longer continue ... But for this simple process, the Australian woods had probably contained as thick a jungle as those of New Zealand or America, instead of the open forests in which the white men now find grass for their cattle ... The omission of the annual periodical burning by natives, of the grass and young saplings, has already produced in the open forest lands nearest to Sydney, thick forests of young trees, where, formerly, a man might gallop without impediment. (Quoted in Rolls 1981, 249)

Aboriginal people managed their Countries to sustain productivity, and that productivity became the foundation of European settlement and prosperity. The rapacious and destructive practices in many parts of Australia, however, meant that settlers had to move on once they had consumed the productivity they had appropriated. From New South Wales, cattlemen took their cattle north into Queensland, and from Queensland they moved west into the Northern Territory and the Kimberley region of Western Australia.

In a recent and constructively critical study of settler Australians' land use practices, William Lines describes the pastoralists' trek across vast portions of Australia as a wave of devastation:

> Heavy stocking and the practice of annual burning off led to a rapid deterioration of pasture. Stock ate the country bare, leaving the exposed soil vulnerable to erosion. Only the hardy, poorer type of vegetation survived ... Those who came in the 1860s land rush [in Queensland] took up run after run, worked feverishly for quick returns, and abandoned each degraded pasture for virgin land further out. (Lines 1991, 106)

Lines notes as well that much land was taken up in speculation, and that absentee owners pushed stocking numbers way beyond carrying capacity (Lines 1991, 106). By 1890 there was an estimated 100 million sheep and nearly 8 million cattle grazing over a huge array of ecosystems that were becoming diminished and impoverished (124).

Almost three-quarters of Australia's land mass is rangeland. A wide variety of environmental regions are rangelands, including the semi-arid inland areas of Queensland and Western Australia, the floodplains, savannas and deserts of the Northern Territory, the inland plains and subtropical areas of New South Wales, and the high mountain plains of New South Wales and Victoria. With low population densities of humans and much higher populations of grazing animals, it is cause for extreme concern that at least one-third of the rangeland area suffers from land degradation. Productivity of grazing animals is in decline, and the loss of native plants and animals is severe. A report prepared by scientists of the Commonwealth Scientific and Industrial Research Organisation (CSIRO) on the future of the rangelands in Australia indicates that nearly half the rangelands' original native mammals (some 72 species) are no longer found in rangelands. Eleven species have become extinct, and 20 species survive only in isolated pockets. In any one region, up to 60 per cent of the mammals that were there before colonisation are now gone (Beale and Fray 1990, 98).

The rate of mammalian extinctions in Central Australia is reported to be the highest in the world.[8] CSIRO scientists ascribe this ocean of loss to three main factors: introduced grazing animals (and overgrazing), feral animals (particularly rabbits, cats and foxes), and cessation of Aboriginal burning (Morton 1990). I will return to issues of burning and other practices of care in Chapter 7.

Cattle are not the only species implicated in these changes. In North Australia, water buffalo, pigs, donkeys, camels, horses and cats have all had their impacts. Of these major species of introduced domesticates, all have gone feral. In the area under study, only pigs have not thrived.[9] Camels are rare, but donkeys have gone feral in pest proportions. Programs to eradicate feral donkeys have been undertaken at regular intervals, and another one (c. 2000) is currently in process. In 1970 the Australian Government

---

8   The precise source for Debbie's statement has not been located, but similar assessments have been made within the past decade. See e.g. Woinarski, Burbridge and Harrison (2012); Stebbings (n.d.)—eds.
9   Based on personal observation by Darrell Lewis, since Debbie wrote this statement (2003) pigs have spread throughout large areas of the district—eds.

initiated the 'Brucellosis and Tuberculosis Eradication Campaign' to clean up all Australian herds in order to bring meat exports up to world standards. North Australia posed particular difficulties and costs of disease eradication were correspondingly high. The campaign offered the opportunity to eliminate feral bovines across much of the north. Cattle were eradicated where they could not be controlled, wild buffalo were completely eradicated over huge areas, and compensation was paid to pastoral lessees (Stoneham and Johnston 1987).

Control of cattle has required more fencing, more bores and dams, more imported hay and even some local cultivation of hay. Each measure of control involves an intensification of land and water use, and each intensification triggers more loss.

3

Walkabout

Bio-eco maps

My early research led me to wonder about the plants and animals people told me about and that we never seemed to see. Ethnobotany was a way of learning more about the plant world, and it turned out that memory was essential. The research into ethnobotany led to a more focused examination of ecological knowledge in general, and to questions of ethics. The research meshed with Victoria River people's continuing struggle for social justice. When I began discussing with some of my Aboriginal teachers the idea of undertaking more thorough documentation of ecological knowledge, Hobbles Danaiyarri approved the idea and said: 'Before *kartiya* [whitefellas], blackfellas bin just walking around organising the Country.' Like so many of his statements, this one carried embedded within it an implicit critique of colonisation. *Kartiya,* he implied, have been walking around disorganising the Country ever since they came here. In subsequent chapters I will explore concepts and patterns of organisation and will look at disorganisation. In this chapter I take up the issue of 'walking around'.

The ethnobotanical work and my previous ethnographic study had enabled me to draw up lists of plants. I wanted to make a concerted effort to locate plants that I had heard about for years but had never seen. Many of the plant species on which people used to rely are no longer within the range of motor vehicles or day walks. Most middle-aged and elderly people had a great deal of knowledge and understanding which could not be demonstrated because there were no accessible environments.

Along with ethnobotanical surveys, I decided to work with maps—to take memory journeys and revisit the places people had walked in their youth. I sought to document the foods they had eaten as they travelled, as this seemed an appropriate way to engage memory with plants, animals and place. Aboriginal Australians generally have superbly developed spatial memory, and stories of travelling, stopping and getting food constitute the structure and much of the content of Dreaming stories as well as personal and familial histories. Inspired by Hugh Brody's (1981) study *Maps and Dreams*, and by Paul Stevenson's work with the Tiwi people of Bathurst and Melville Islands (1985), I decided to map people's travels, recording the places they went and the food resources they used.

Between 1986 and 1991, I carried out biographical-ecological mapping with 15 people who, if they were all alive today, would range in age from over 100 years to about 60 years. Ten were men, five were women. The maps document resources in some areas for which no scientific surveys have been conducted. In addition, they document distributions from a time which is no longer accessible. In accessible areas such as nearby billabongs, the maps document former abundance of some species which are now absent entirely, or present only sporadically and in scarce amounts.

I took an individual approach, conducting interviews with one person at a time, and encouraging others to save their stories for their own map. Usually, a group of people listened to the interview, and sometimes others commented on the stories offered by the individual who was the primary speaker. I limited the maps to footwalking; droving trips and mustering trips were excluded for the purposes of this study. Accordingly, I worked with people who had been born early enough to have become young adults before motor vehicles became widespread. I worked with 1:500,000 scale maps as I knew from experience that smaller scale maps were still not accurate enough to provide the level of detail that would allow one to retrace a particular trip step by step, spring by spring. Using inexpensive dye-line maps, I wrote directly onto the map, and taped the conversation. These interviews produced vast amounts of information above and beyond documenting travels and food, but in no way did they explore the full range of the participants' knowledge. Some people found the process of listing foods tedious and sped it up by lumping foods together (plum, instead of several different kinds of plums; goanna instead of different kinds of goannas). More significantly, however, I was asking primarily about food, as I had a list of foods for which no specimens had been found. My lists did not include any medicines or technological items that had not been identified, and therefore I did not probe the locations of medicines and technological resources.

The interviews gave pleasure both to the participants and to the audience group. Every person who demonstrated a breadth and depth of knowledge of Country, resources and skills expressed pride in that fact, and we all became interested in finding out who had 'the biggest map'.

The colonising regime

All the participants in this study were born into relationships with Country, with other species, with other humans, with song cycles and ceremony tracks, Dreamings, designs and with places of power, as well as with foods and water sources. They were born for Country, in short, and nothing that had happened in the course of their lives had altered these foundational relationships. They were, however, wards of the state and their human rights were severely restricted until the mid-1960s.

Colonising society had laid a grid of cattle stations over their Countries, and so they were born into relationships with these vast properties. They were ostensibly under the protection of the 'crown', and that relationship was mediated by the Protector of Aborigines (later the Department of Native Affairs, and the Department of Aboriginal Affairs). Locally, however, they were answerable to the police, the station manager and the station personnel.

Most of these people worked for Wave Hill, Humbert River and Victoria River Downs stations. They learned the Law for their own Countries and regions; they learned the detailed knowledge of Country that enabled them to keep themselves and their families alive when station rations were at starvation level or non-existent. During the 1960s and 1970s they walked off the cattle stations, withdrawing their labour and demanding land and justice. Following the walk-off they went to live in communities where they were, for some time, out from under the strong scrutiny of white authorities: Yarralin, Lingara, Daguragu.[1]

---

1   The Pigeon Hole people returned to the same community, immediately adjacent to the homestead —eds.

During the working year, 'rations' were supplied to workers and dependants under a scheme whereby stations received the labour of the workers in exchange for food, which was subsidised by the Australian Government (Stevens 1974, 82, 91; see also Rose 1991). Reports by committees of investigation from early to late indicated widespread under-resourcing of rations. One effect of insufficient and inadequate food was to require people regularly throughout the year to rely on bush tucker as well as rations. One effect of sedentarised use of bush tucker was over-utilisation of resource sites that were within walking distance of the station homesteads where people lived for much of the year.

There were positive benefits to the reliance on bush tucker: people had a varied and more healthy diet than would have been the case if they had lived on rations alone (flour, sugar, tea and beef were the staples). Knowledge of foods and food processing techniques were kept alive. Children carried out much of the food gathering, working to feed their parents who were required in station work. Jessie Wirrpa described childhood walkabout with pleasure:

> We used to, all the kids, we used to go down to the river and getem that mussel, make a dinner camp, boilem that mussel, and *yawu* [fish, generic], catfish or bream. Silver fish, little *danyan,* you know. We used to boilem that mussel, oh, big mob. Takem back for mother, family.
>
> Debbie: This just all you kids been go walkabout?
>
> Jessie: Yeah. All go down the big river, getembad mussels, all the kids. Longa Jirrikit [Dreaming site for her, Jirrikit Dreaming, discussed in Chapter 4] … We used to [be] fishing all day, catchembad *yawu,* and afternoon we go back home, takem *yawu,* and mussels for mother and granny.[2]

There were also negative effects, as I have mentioned: long-term degradation of resource sites, and widespread malnutrition at various times and places (discussed in Rose 1991; see also Berndt and Berndt 1987 for an excellent discussion of the social factors of malnutrition and starvation in the 1940s).

The practice in the Victoria River District was to work the Aboriginal labour force during the mustering season (roughly, the dry season) and then to release them from labour and encourage them to go bush during the rainy period.

---

2   Jessie Wirrpa, tape 78, recorded at Yarralin, 13 July 1986.

Encouragement took the form of giving people a few rations and telling them to come back when it was time to muster again. People really had no choice but to go bush where they could get food to keep themselves alive during the time away from the station. In addition, the brutality of station life pushed people into the bush for longer or shorter periods of time (see Snowy, below).

Numerous scholars have noted that this period of enforced walkabout was a critical factor in people's efforts to sustain their relationships with Country, their knowledge of Country, and their religious and social practices. It must also be noted, however, that the rainy time of year is a time when travel is difficult and when resources are not yet replenished by the rains. Prior to colonisation, large inter-group gatherings took place just after the rains when travel was becoming easier, and many foods were in abundance. One of the effects of colonisation, therefore, was to shift the intertribal ceremonies from a period of plenty to a period of relative difficulty both in terms of available foods and in terms of mobility (see Dora, below).

In spite of the periodic hardships of the past, the participants in the map study spoke fondly and enthusiastically of bush tucker, and characterised today's reliance on money and *kartiya* food as negative factors in a number of dimensions (see Kitty and Hobbles, below).

Life histories

I present portions of four life histories; portions of many others are included throughout this book. I have not sought to include lists of foods in their words, or to describe every trip; rather I have extracted some of the most characteristic and some of the most distinctive parts of each interview. I distil the list of all foods mentioned by Snowy later. The main themes that arose in these life history interviews are happiness and pride, expressed through knowledge and nurturance. My purpose in presenting these texts is to enable people's own words to communicate in their own way.

As I will discuss in detail in Chapter 4, my teachers divide the Victoria River valley into a number of ecological zones, of which three figure prominently in their lives: savanna desert Country, river (freshwater) Country and saltwater (tidal influence) Country. The home Country of three of the four teachers whose narratives are excerpted here is in river Country; the fourth is in desert Country. Three of the four speakers tell of their encounters with ecological difference. Their stories thus give us their own insight into the warmth of home place, and the recognition of difference.

Dora Jilpngarri

Dora Jilpngarri was about 80 years old in 1986 when we made her footwalk map. She had been a valued employee of Victoria River Downs (VRD) station and spent most of her life at the head station (Centre Camp).

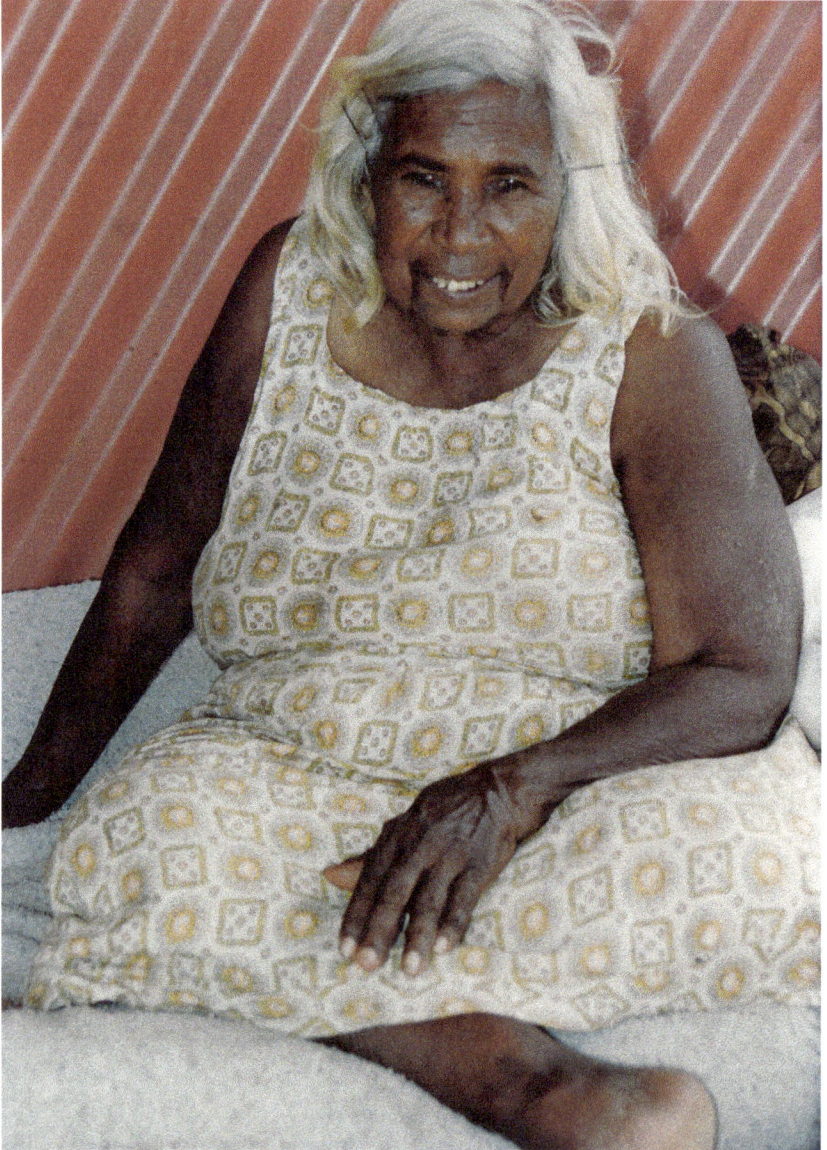

Figure 3.1. Dora Jilpngarri, Yarralin, 1982.
Source: Photograph by Darrell Lewis.

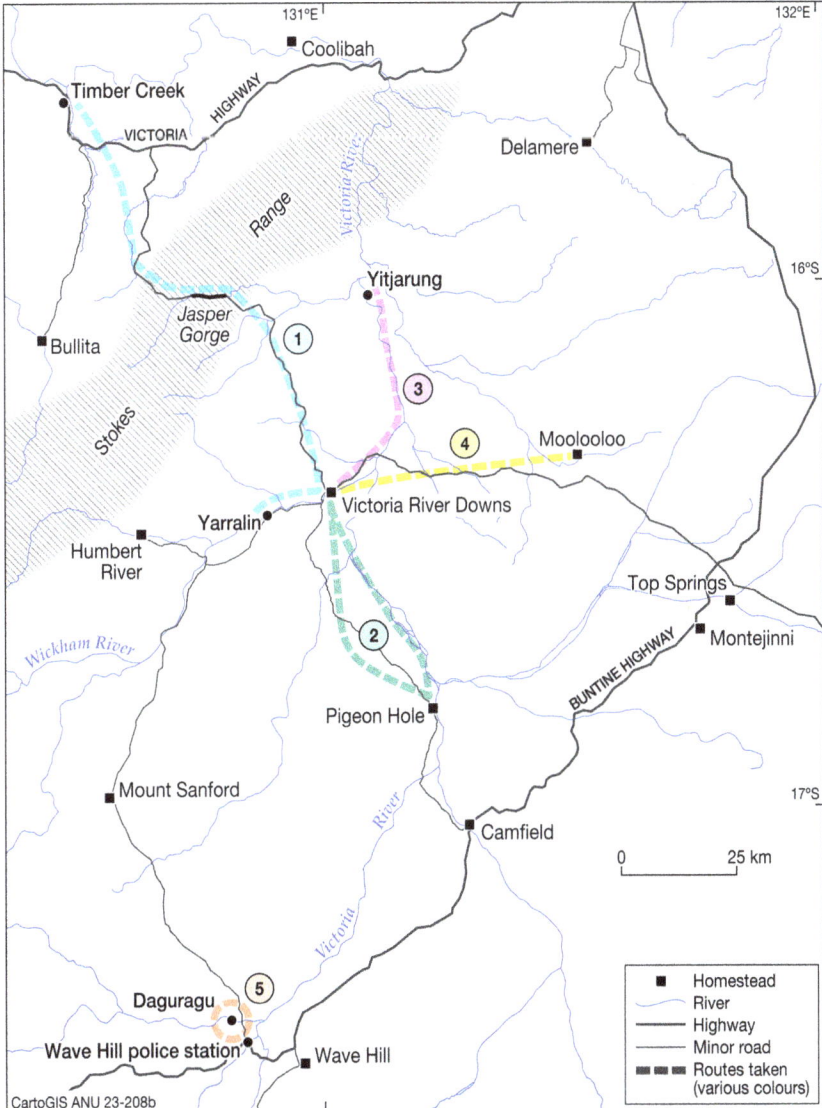

Map 3.1. Dora Jilpngarri's footwalk map.

Source: Karina Pelling of CartoGIS ANU.

1.	*Jalyingarna* (sugarleaf — insect exudate on bloodwood), *janaka* (native rosella — inner of stalk eaten), *jikamuru* (water lily), *kakawuli* (long yam), *kamara* (black-soil long yam), *kilipi* (bush banana), *kitawa* (goanna), *kumpulyu* (white currant), *markul* (like a water lily), *mintarayij* (water lily), *muyin* (black plum), *namawurru* (tree sugarbag), *nankalin* (ground sugarbag), *ngamanpurru* (conkerberry), *pumparta* (lily seeds), *pampilyi* (vine berry), *purrungurn* (sugarleaf — insect exudate on river red gum), *tipil* (black currant), *walmatj* (*Livistona* palm), *wayita* (small tuber), *wilit* (small black fruit and spear shafts), *yamali* (fruit like a mango), *yarkalayin* (hairy yam).

2.	*Japungarna* (water goanna), *kitawa* (goanna), *namawurru* (tree sugarbag), *warritja* (freshwater crocodile), *wayita* (small tuber), *yarkalayin* (hairy yam), *yawu* (fish).
3.	*Jiya* (kangaroo), *kayalarin* (bush onion), *kitawa* (goanna), *namawurru* (tree sugarbag), *yawu* (fish).
4.	*Danyan* (spangled perch), *karayij karayij* (aka *martarku*, green plum), *partiki* (nutwood tree/seeds), *tuku* (freshwater mussel), *yawu* (fish).
5.	'Strike time' [this is when Dora was temporarily living at Daguragu], catfish (*warak* or *jalarlga*), *kuwalampara* (turtle), *warritja* (freshwater crocodile), *warritja kambij* (crocodile eggs), *yawu* (fish).

I bin borning longa VRD. I bin born longa VRD. We never go any way. I bin born there, longa VRD. When I bin little bit big now, well mother bin take me walkabout longa bush. Walkabout longa bush this way, [to the] river. Any way, im bin takembad me. Havem *wayita* [little tuber], havem *jikamuru* [water lily], all right. We bin go longa [eat] *muyin* [bush plum] and *tipil* [black currant, *Antidesma parvifolium*], where im bin give me, growem up. My mother bin only feedem up me la bush. All right. Eatem *kilipi* [bush banana], and we blackfellow call im *yamali,* like a mango, you know. Different way, you know. Like a mango. We bin always, I bin always havem *namawurru* [tree sugarbag]. And sugarleaf.

Sugarleaf is an insect exudate that appears on the leaves of two different eucalypts, one riverside and one topside. The two different eucalypts produce different flavours and textures of sugarleaf, and each is prized. Sugarleaf is a cold weather food and is harvested and processed by women. Dora continued:

> *Jalyingarna* [one kind of sugarleaf]: im good. And sugarleaf, nother sugarleaf *purrungurn.* Like a porridge, im like a porridge. Longa riverside, no more top—riverside. He sit up like a, he's sugarleaf, only he's different, like a porridge. Sweet one. Same [like] porridge, like that now. He's good tucker, all the same. Im only [found along] river side. Put it in a bag, and im good. Take im away. Wet im, and make im round one like [a cake], and eat im, any way.
>
> *Jalyingarna* longa bloodwood leaf, and nother sugarleaf now, im longa *malan* [river red gum] ... Oh, we bin always get a big mob there when I bin little

girl. I bin big girl too you know. Like that time. I bin
always getembad [get sugarbag] meself when I bin big
girl now. I bin always getembad meself. I bin do it, all
the same. Mother bin always showem me, you know.

In speaking of childhood Dora kept returning to her mother's nurturance.

And nother one, nother one like a plum, *kumpulyu*
[white currant]. And *kumpulyu,* Mother bin always
feedem up me when I was little.

Memories of her mother triggered memories of sugarbag, the honey
produced by native bees. It is classed according to where the bees make
their hives: in trees, in the ground (under rocks), or in termite mounds.
Dora's discussion of sugarbag brings together nurturance, knowledge and
experience:

And mother bin always give it me sugarbag [native
honey] from tree, mother bin always give it me. And
nankalin, ground sugarbag. *Nankalin,* we call im,
where im sit down in the ground. Good one too,
eh? He can't make you sick, from ground, you know,
sugarbag. He's good one properly, he can't make you
sick. No mater how much you eat im, he can't make
you sick. He's good one. Sugarbag longa tree, he'll
make you sick when you havem too much, you know.
Im makem you sick. No more that one longa ground.
He's good one. He can't make you sick. Clean one too.

Along with her mother were all the old women, now passed away:

When I been longa VRD we been go holiday here,
climb up la this Parlimulmul [on top of the Stokes
Range]. Umbrella house, eh. Umbrella house
[a massive boulder shaped like a mushroom, with
a wide overhang all round]. We been camping out
there, now we bin go that way, we been go that way,
camp longa that nother place, right, we been go camp
longa that nother place, might be Jilatmani, that one.
All right, we been come back now. We been come back
for go down long this Deep Creek. Right, that, that
nother old woman been come up and tellem mefellow,
'Come on, let's go gettem that tucker. Big mob there.'
Oh, we been gettem, fillem up longa all the billy cans,
you know. Just washem and eatem. We been bringem
down right up longa Deep Creek, that tucker.

Dora's father also nurtured her:

> Ah, *makaliwan* [wallaby]. *Makaliwan.* My father bin always kill im longa [for] me. And rock kangaroo. Big one. *Jiya* now. *Jiya*—we call im that one.

Dora described travelling north from VRD Centre Camp to a waterhole called Larry's Lake. This was a regular trip for her family, and the retelling brought back memories of plenitude.

> When we bin go Larry Lake, oh, we bin always eat im goanna, kangaroo. We bin always eat im all the way, go longa river, findem fish now, *yawu* [fish] now findem. Sit down longa river. Sit down longa *yawu.* Ah, sugarbag, we bin always find im. We bin always find im. We bin go, when I bin little girl like that, you know, we bin go right up. We bin right up there when I was little girl. My father bin always take me. And mother. Mother bin always taking me walkabout. We never hungry when we bin kids, you know. We never hungry. Mother and father bin always feedem we. We never bin hungry, little ones, little girl. They always feedem [us] up.

One of Dora's most memorable trips was the time she and her husband Jimmy, along with a larger group, walked, swimming across the rivers, to Coolibah Station for 'business'. Coolibah is about 120 kilometres as the crow flies from VRD Centre Camp and is beyond the riverine zone in which Dora's home Country is located. The group travelled through Jasper Gorge in order to get to Coolibah, and on the way they were caught by huge monsoonal rains. Having gotten beyond the range of Dora's and Jimmy's own knowledge, they had to rely on others in the group to teach them about the Country and different resources.

> And going to Coolibah, that's [when I was] the big girl now, when I bin go married now. And me and Jimmy bin going there now. Long way that Coolibah. We bin going up the road [for] four months. Footwalk. Four months.

> Too much wet weather, you know, rain bin all day, all day, all day, all day. We never catch em up quick. Four months longa road.

> Debbie: Up through *Kuwang* [Stokes Range]?

Dora: We bin cut across, you know. Longa that road, here [to] Potato Spring. Potato Spring this way, going up to Timber Creek … Well after Potato Spring now, we bin go cut across now. Leavem that road. We bin going bush [cross-Country] now. Blackfellow road we bin followem. Blackfellow road. We bin go Jasper Gorge. We bin go from here [to] Jasper Gorge.

Figure 3.2. The Black-Headed Python Dreaming boab at Manjajku (Bottle Tree Waterhole), Jasper Gorge, 1982.

Source: Photograph by Darrell Lewis.

We bin go from here right up to Gorge, longa that big bottle tree. We bin go down there. We bin find em that *walmatj* [palm tree, tucker], that *walmatj* we call im. They bin always break im, clean im up, they bin always give it we. We no more savvy that *walmatj*. We no more savvy eating *walmatj* before. We don't know. We bin try im—oh, im sweet! … we bin likem eat im. We bin [keep] going up. We bin always findem little bit of *wayita* [yam] there longa road. That big *wayita,* you know, big one *wayita.* Proper *wayita.* Only big one. Different from this place [VRD], this one. That's big one *wayita. Minterpala* [we, inclusive] bin kill [dig up/cook] im and big mob, well all bin go, camp. Killem little bit of *kakawuli* [long yam] there, *kakawuli* there, big mob *kakawuli.* We bin killembad all the little ones, you know. No more big ones, all the little ones. Little ones all the *kakawuli* we bin killembad. Eat im.

All right, next time we bin getem *markul. Markul* [a kind of water lily] there little bit of billabong there gotem *markul.* That side billabong, we bin getem *markul* there, little bit. All right. We bin eatem all the way sugarbag [honey] now, cutembad [chop out] tree sugarbag, all the way. Go longa that Potato Spring now. Nother *kakawuli* they bin find em. They bin give it we [to us] for eating, *kakawuli.* We bin camp there two nights longa Potato Spring. All right. Tomorrow morning we bin cross, now, leavem that road now, main road. We bin go longa blackfellow road, now, cut across. Findem all the way sugarbag, that's all. Findem sugarbag, one each. Sometimes goanna, findem. Next time nothing. Next time findem goanna. All the way like that. Too much rain. You know, too much big rain bin going up all the way. We never catchem up quick that Coolibah. We bin long time longa bush. Long time. Never catch im up.

Belong to little boy business [initiation]. We bin takem two little boys. We bin corroboree there long Coolibah. Where we bin come back, that's the one plenty tucker. Lotta *kakawuli* killembad [digging up and cooking] all the way *kakawuli.* Killembad all the way. We bin cross now, go longa Timber Creek. We bin follow that nother road now. Timber Creek way, now, we bin followem. We bin havem *nankalin* [ground sugarbag], big mob. *Nankalin* big mob … Only we bin [eating] *kakawuli* now, come back time … No more findem more beef [meat] now, only *kakawuli* we bin eatem when we bin come back.[3]

Snowy Kulmilya

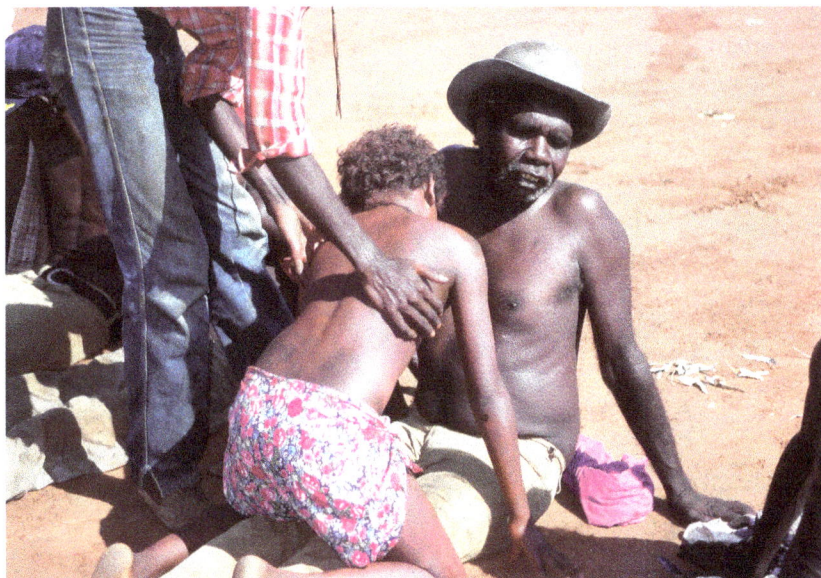

Figure 3.3. Snowy Kulmilya engaged in young men's business, Yarralin, late 1981.

Source: Photograph by Darrell Lewis.

---

3   Dora Jilpngarri, tape 82, recorded at Yarralin, 18 July 1986.

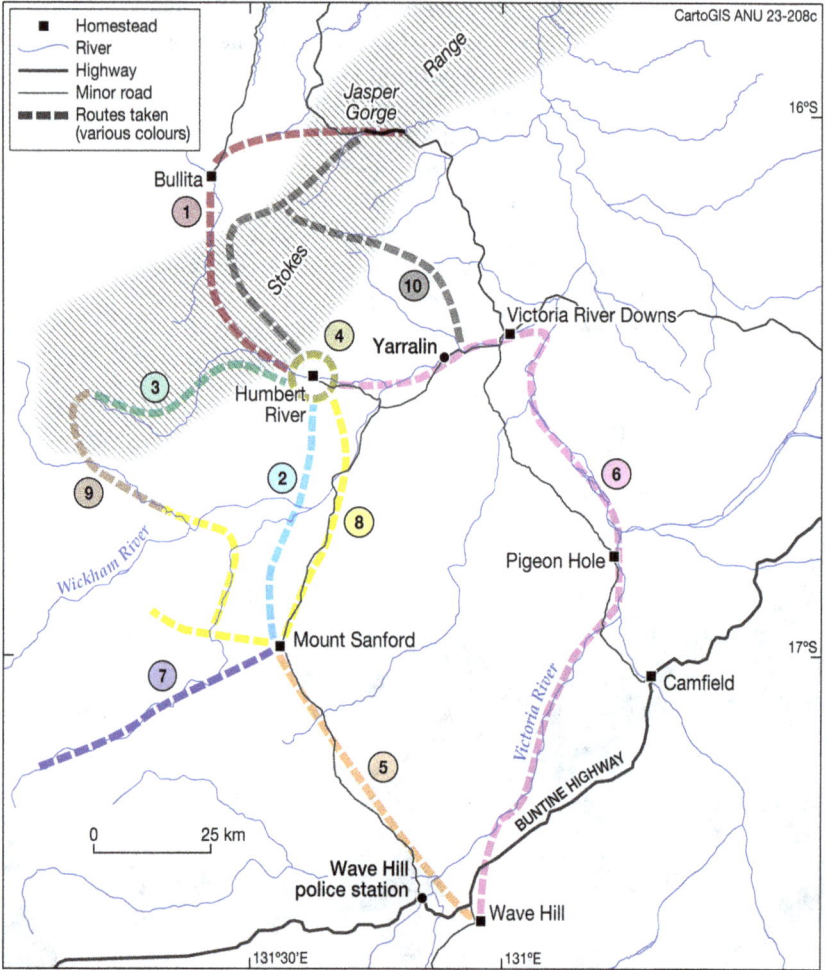

Map 3.2. Snowy Kulmilya's footwalk map.

Source: Karina Pelling of CartoGIS ANU.

1.	*Didi* (peewee), eaglehawk (wedge-tail eagle), *jakarin* (sand Country, like a potato), *jalyingarna* (sugarleaf on bloodwood), *jamunang* (gooseberry), *jamut* (plains turkey), *jangana* and *jarkulaji* (brush-tail possum), *japungarna* (water goanna), *jarrwana* (*Pandanus aquaticus*), *jibilyugu* (duck), *jikamuru* (water lily), *jimanik* (unidentified), *jimilawumuru* (unidentified), *jiya* (kangaroo), *junkuwuru* ('porcupine' echidna), *karil* (wild cucumber), *karrang karang* (diver duck), *kayalarin* (bush onion), *kitawa* (goanna), *kitikiting* (like a water goanna), *kitpan* (bitter cucumber), *kulijpa* (yellow kapok), *kulpin* [*kulpun?*] (hawk), *kumira* (budgerigar), *kunamulun* (olive python '*kunutjari*'), *kurangij* (unidentified), *kuwalampara* (turtle), *lamawut* (witchetty grub), *lamparlampara* (bush tobacco), *malajaku* (big sand goanna), *maran* (wild pussy cat '*walamunpa*'), *marpalangpalang* (pigeon), *mijat* (unidentified), *mintarayij* (water lily), *nankalin* (ground sugarbag), *ngamanpurru* (conkerberry), *palatmawu* (bandicoot), *parangara* (white cockatoo), *partiki* (nutwood tree/seeds), *pulkal* (black plum), *purrungurn* (sugarleaf from river red gum), *tilipi* (possibly fig), *wajipat* (bush potato), *wajilarn* (galah), *wak wak* (crow), *walmalmaji palarr* (stone Country tobacco), *walujapi* (black-headed python), *walukpil* (black cockatoo), *warritja* (freshwater crocodile), *warrpa* (flying fox), *wayita* (small tuber), *wititpiru* (spinifex pigeon), *yaramulku* (tree sugarbag), *yarkalayin* (hairy yam), *yawu* (fish), *yiparatur* (emu).
2.	Same tucker as 1.
3.	Same tucker as 1 and 2, including: *kamankira* (spear grass), *kitawa* (goanna), *kungkala* (fire-stick), *kuwalampara* (turtle), *namawurru* and *nakalin* (tree and ground sugarbag), *pulkuru* (bamboo?), *warritja* (freshwater crocodile), *wilit* (small black fruit and spear shafts).
4.	*Kakawuli* (long yam), *pikurta* (bush potato).
5.	*Jiya* (kangaroo), *kayalarin* (bush onion), *kitawa* (goanna), *kuwalampara* (turtle), *mintarayij* (water lily), *takirin* (Polynesian arrowroot), *warritja* (freshwater crocodile), *yawu* (fish).
6.	Same tucker, including: *kuwalampara* (turtle), *namawurru* (sugarbag), *wayita* (small tuber), *warritja* (crocodile), *yawu* (fish).
7.	Same tucker as 5 and 6, including: *japungarna* (water goanna), *jikamuru* (water lily), *jiya* (kangaroo), *ngamanpurru* (conkerberry), *namawurru* (tree sugarbag), *warritja* (freshwater crocodile).
8.	[notes indecipherable].
9.	[indecipherable].
10.	Rock holes and springs (Mulutpayi, Kajutarnang, Jangara, Kulinjiwuru), antbed sugarbag.

Snowy Kulmilya was born on Humbert River Station in about 1927 and grew up there. As he said, 'I'm belong to that Country.'[4] His footwalk experience was only moderately extensive compared to others, but it was remarkably intensive. He travelled with his father in areas of his father's Country that were difficult of access other than by foot, and he came to know them very well.

---

4   Snowy Kulmilya, tape 89, recorded at Yarralin, 27 July 1986.

When I bin kid, when my father was alive, he bin take me walk, sometimes months, four months, two months, like that, la bush. Come back. Just living on bush tucker.

We used to run in the bush, no tobacco, no tea, no sugar, no nothing, no [white man's] tucker. We used to living longa lily and all the bush tucker ... Or kangaroo. We never run out for tobacco. Plenty of bush tobacco la bush, you know. And tea. That *lamparlampar* [bush tea], you know. Put im la bag, fill im up. Make it ... every morning. And that *jikamuru* [water lily] good enough, [and] sugarbag. We never think about all them *kartiya* tucker, you know. We used to go la bush. That's where I learned. Just living la bush tucker.

He described walking from Humbert River to VRD:

Goanna. We bin living [on] goanna, turtle, crocodile, and fish. Kangaroo sometimes. When we couldn't findem turtle, anything, la river Country, when we too late, well we just see kangaroo on the road, well we just kill im. That's all. Kangaroo, turtle and fish, and crocodile. And sometimes we bin go longa billabong getem that lily. *Mintarayij* [water lily] now.

Like many of the speakers, Snowy spoke of the toxic tuber *kayalarin* which must be prepared according to a long and arduous procedure in order to make it edible.

And that *kayalarin,* that cheeky one. We bin getem that one, some old old women, they bin always getem, cookem up, and getem stone, smashem up and makem properly black. When you eat that half cooked, he's burn all your mouth. No good. Pretty good tucker when im cookem up properly. Good one. You can eat.

Snowy spent time living in the bush, not just at holiday time but for longer periods as well, to escape the hardships of station life. (For more about life on Humbert River, see Rose 1991, 229–35, 244–45.)

Oh, we didn't want to run away. But that head stockman didn't, you know, we didn't like him hurting us. We didn't like it ... we were too small. And, that was early days. We all bin frighten. Them old people didn't take a place [stand up] for us too. You know, poor buggers, some of them reckoned, 'No good. Never mind. You'll have to go away tonight and see your father out bush. Them should be this way la so and so place.' … Soon as that sun go down, just roll our swag and go away …

Debbie: So when you were a kid, young fellow, anyhow, there were people living in the bush, you could run away easy findem?

Snowy: Yeah. We used to go down there and find them. We know where that place where they always live. We know where the spring place. We used to go down there … Just follow the track, from old track to new track. We knew where he's living. We know he's just going straight in that place.

When Snowy ran away to be with bush people, they were living in relatively inaccessible areas: the rough Country high up on the stony watersheds between the big rivers. This is Country where the locations of water were not self-evident and where knowledge made the difference between life and death. Much of his map shows a deeply detailed knowledge of the locations of water, and his account of time in the bush names one spring, rock hole or crevice after another. He described a rock hole within which the water was so deep that even birds could not access it.

You'll see flat rock there … He got a hole like that, not much wide … And water there inside. You got a bit of string, billy can on a string. Let em down, get that water there. No bird can sing or anything. Water is there inside. [It is a] well. We used to get the water and chuck im on a flat rock there. Feed [give water to] them dogs, feed them up.

The dogs helped with hunting: 'They always chase kangaroo … They bin always hunting, them dogs, and chasem them and catchem kangaroo too.'

Snowy and others reciprocated, sharing food with their dogs.

Snowy explained the use of a fish poison made from the leaves and pods of *parrawi*. When the leaves and pods are wetted and rubbed, they make a lather that poisons fish. This method is used in small, contained water sources where the poison will work without being washed away:

Nother spring callem Karlayi. Karlayi, im right up la hill. We bin getem that *parrawi*. You know *parrawi*, that *parrawi* tree. Getem that leaf and killem that leaf

now, makem that soap. Killem all the fish. Get all the
fish there. But we didn't frighten [for] drinking that
water [ourselves]. We bin always drink. Yeah. When
he settle down, you can still drink. He can knock that
fish, but you right for drink.

We bin get them dead, [and] you see em them small
one [fish], chuckem out outside there [from] that
water. Give it la them dog to eat. You know that little
one little one fish.

Snowy claims his history and his knowledge as an outcome of footwalk and
the teaching of his Elders. Both are defining features of who he is and where
he belongs:

I know that far when I bin running la mine father
and mother. I bin learning all the time. That's what
I know tucker from here properly, and beef, I know
what I can eat im from bush. I know which way I'm
going to. I know the Country.[5]

Kitty Lariyari

**Figure 3.4. Kitty Lariyari (left), Maggie John (middle) and author (right),
Yarralin, 1982.**

Source: Photograph by Darrell Lewis.

---

5   Snowy Kulmilya, tape 90, recorded at Yarralin, 27 July 1986.

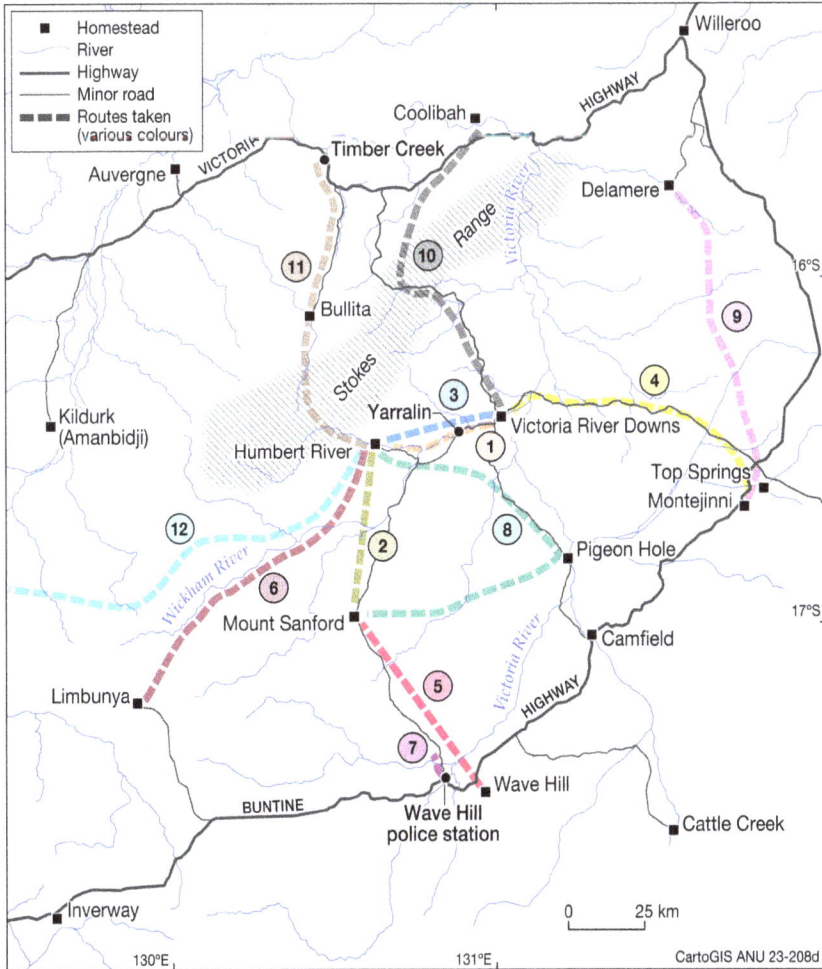

Map 3.3. Kitty Lariyari's footwalk map.

Source: Karina Pelling of CartoGIS ANU.

1.	*Jikamuru* (water lily), *jiya* (kangaroo), *junkuwuru* ('porcupine', echidna), *kitawa* (goanna), *namawurru* (tree sugarbag), *ngamanpurru* (conkerberry), *kuwalampara* (turtle), *warritja* (freshwater crocodile), *wayita* (small tuber), *yawu* (fish).
2.	*Jarkulaji* (brush-tail possum), *japungarna* (water goanna), *jarrwana* (round yam), *jikamuru* (water lily), *junkuwuru* ('porcupine', echidna), *kanjalu* ('water chestnut'), *karil* (wild cucumber), *kartkarta* (*Gomphrena canescens*, use not recorded), *lunkura* (blue-tongue lizard), *mangurlu* (seeds), *nankalin* (ground sugarbag), *wayita* (small tuber), *yarkalayin* (hairy yam), *yawu* (fish), *yingki* (kurrajong seeds).
3.	*Jamut* (plains turkey), *jikamuru* (water lily), *jiya* (kangaroo), *kitawa* (goanna), *karil* (wild cucumber), *muyin* (black plum), *namawurru* (tree sugarbag), *ngamanpurru* (conkerberry), *yawu* (fish).

101

4.	*Jiya* (kangaroo), *jikamuru* (water lily), *kitawa* (goanna), *muyin* (black plum), *namawurru* (tree sugarbag), *pararayij* (green plum), *wayita* (small tuber).
5.	Same tucker as elsewhere, including: *jiya* (kangaroo), *karil* (wild cucumber), *kitawa* (goanna), *muyin* (black plum), *namawurru* (tree sugarbag).
6.	Same tucker, no different from elsewhere, including: *jikamuru* (water lily), *kamara* (black-soil long yam), *walmat* (*Livistona* palm), *wayita* (small tuber), *yarkalayin* (hairy yam).
7.	Same tucker as elsewhere, including: *jikamuru* (water lily), *karil* (wild cucumber), *kuwalampara* (turtle), *muyin* (black plum), *namawurru* (tree sugarbag), *waiyita* (small tuber), *yawu* (fish).
8.	*Jiya* (kangaroo), *kamara* (black-soil long yam), *kirrawa* (goanna), *muyin* (black plum), *nalja* (sand goanna), *namawurru* (tree sugarbag), *walaku* (dingo), *wayita* (small tuber), yams (not specified).
9.	Same tucker as elsewhere, no different, including: *muyin* (black plum), *wayita* (small tuber), no *kamara*, no *kakawuli*.
10.	Same tucker, including: *kakawuli* (long yam), *karil* (wild cucumber), *markarin* (bush grape), *muyin* (black plum), *pararayij* (green plum).
11.	*Jurulana* (small tree with fruit), *kakawuli* (long yam), *karil* (wild cucumber), *kumpulyu* (white currant), *muyin* (black plum), *pararayij* (green plum), *tipil* (black currant).
12.	Same tucker, including: *jikamaru* (water lily), *walaku* or *nurrakin* (dingo, dingo pups), [the remaining names are indecipherable due to damage (— eds)].

Kitty Lariyari spent most of her life working for Humbert River and other stations, and many of her travels are interwoven with her working life.[6] Lariyari was about 67 in 1986 when we did this interview. Her sister Kitty Maliwa was a keen listener. Maliwa had never been able to travel much because as a child she was crippled. She provided commentary on her sister's story.

> Lariyari: I bin borning longa VRD. My daddy bin working, mother mine, they bin working la VRD. Im bin engineer, mine daddy. I bin there until I bin big girl. Mine uncle bin takem me for married now, la Humbert River.[7]

> Debbie: Well when you were living at VRD, when you were little girl, what tucker you bin havem that Country?

---

6 Kitty Lariyari was also known as Dadada—eds.
7 Kitty's uncle gave her to her promised husband at Humbert River Station.

Lariyari: Tucker? What about lily? *Jikamuru. Jikamuru* and *namawurru* [tree sugarbag]. And *kirrawa* [goanna], and *junkuwuru* [porcupine/echidna], and the *jiya* [kangaroo], we bin havem *dumaj* [lots and lots]. But river country crocodile we bin havem. My daddy bin getembad. *Kuwalampara* [turtle]. And *yawu* [fish]. *Kirrawa, yawu,* im bin always getembad, mine, mine daddy. Mine daddy bin getembad gotem *milarang* [with a spear]. Goanna, and *jiya* im bin killem gotem *milarang.* Im bin takem me walkabout. While im bin walkabout, im bin killem food for me, *jiya, kirrawa, yawu.* Sometimes mine mummy bin going hunting for *jikamuru,* for sugarbag, *namawurru,* im always go hunting. And *wayita* [small tuber].

My husband bin give it me same way. When I bin married. Same way im bin givem me. Im bin takem me la bush too, walkabout. Holiday, you know. La bush. La my Country! Back to Layit [upper Wickham River] and come out la [Mt] Sanford [outstation]. All around Country im bin takem me, when I bin married. But I never bin go hungry.

Mmm. *Jikamuru, wayita, namawurru, kanjalu* [probably a type of 'water chestnut'], *junkuwuru* [porcupine/echidna], *kirrawa* [goanna], *yawu* [fish], blue-tongue [lizard], we bin tuck out. Im bin always feed em up me, makem me growing up, my husband. All around im bin takem me la bush, holiday, you know, I bin always come back la job. We bin working Humbert. I bin horseman, you know, me.

Debbie: You? Riding horses?

Kitty: Yeah. When I bin young.

Next time, mefellow bin always go bush again, holiday. Come back this way now, la VRD. Mine daddy bin live yet. Footwalk. Footwalk job. This time we got a motorcar. No more first time holidays, nothing. We bin always footwalk. (For more of Kitty's working life, see Rose 1991, 203–4.)

Figure 3.5. The Pilimatjaru sandstone, Gordon Creek, VRD, 1984.
Source: Photograph by Darrell Lewis.

Kitty described a route of travel in the holiday time that took her south and west into her father's and father's fathers' Country. They made a big loop from Humbert River Station to Mount Sanford (an outstation of VRD) following the Wickham River for part of the way. From Mount Sanford they travelled through the rough Pilimatjaru sandstone Country to Pigeon Hole and then back to Humbert River.

> Mefellow bin always go Sanford, Pigeon Hole, Wave Hill. From Sanford mefellow bin go this way now, river, all the way, la that Pilimatjaru River, eh? Pilimatjaru River now mefellow bin always go. Oh, mefellow bin findem *kirrawa, wayita, ngamanpurru, jiya, kirrawa,* too much. *Wayita,* we never bin go hungry, nothing. *Jikamuru.* Any kind *mangari* [tucker from plants]. *Muyin* [plum]. You know *pararayij?* [That's the] green plum. Mefellow bin tuck out, right through, come out la Pigeon Hole. From Pigeon Hole, mefellow bin come up this way, Whitewater River [on Humbert River Station], Whitewater come out. Footwalk, you know. Come out la William Yard. Go la Humbert now. Go work now. Likey that all the way.

Another of her trips took her to Limbunya Station. This trip too started out following the Wickham River, which is the defining river of her mob of people. Following it into Limbunya they went into the region of the headwaters.

Wickham River, mefellow bin follow aaaaalll right up. We bin turn off now. Proper bush now, mefellow bin go. Camp half way. From all the way now, come out Limbunya now. *Namawurru* [sugarbag] again. *Muyin* [plum], same tucker. *Namawurru*, oh, you can't beat im *namawurru*. That Country proper [best sugarbag]. *Jikamuru, wayita,* you know *kamara* [yam]? That yam, long long one. That *mangari* too mefellow bin tuck out … I bin all round Country, me. Footwalk.

When I bin losem mine husband, you know. Im bin passed away, well I bin come up from Humbert River. I bin come up Centre Camp now [VRD], I bin working. From Centre Camp … [go to] Montejinni … We bin helping that [Montejinni] missus bela ironing clothes, for washem clothes, and settem table, likey that, you know …

[Tucker], oh, *wayita* [small tuber] again. Same *mangari* [tucker]. Like this way where mefellow bin havem. Same *mangari* we bin havem [at] Montejinni. You can't beat im bela *kirrawa* [goanna], bela *jiya* [kangaroo], Montejinni country. *Kirrawa, yawu* [fish], all the same. *Namawurru* all the same. *Muyin, kamara, wayita.* Montejinni country you know, proper big big one [*wayita*].

Kitty described learning to eat palm heart in Limbunya Country where the *Livistona* species, as yet undescribed, is said by Aboriginal people to differ from the one in Jasper Gorge. At any rate, once she learned to eat palm hearts she ate them as well when she travelled through Jasper Gorge.

Kitty: You know *walmat?*

Debbie: Yeah.

Kitty: Yeah. *Walmat* I bin havem la … Limbunya. Limbunya I bin tuck out *walmat.* Then I bin savvy now for Coolibah country, *walmat.* This one [Jasper] Gorge, eh? Plenty there now, *walmat.*

I bin la Bullita. Just look around, that's all, [for] holiday. Sometimes I bin go gotem dray now, you know, wagon? There now *kakawuli* [long yam] big mob. Big mob *kakawuli* there now, mefellow bin always killembad [dig up/cook] … We never bin hungry la bush …

Kitty's younger sister, Maliwa, asked her: 'You bin havem frog?'

Kitty Lariyari: [In] that Country alabad [everyone] havem. Montejinni country, you know.

Kitty Maliwa: You bin havem?

Lariyari: Yeah. Sand ground country. No more this one river country. [That one] Where im sit down la *kaja* [desert]. Proper good feed.

Maliwa: You like im, Debbie?

Debbie: I had the leg part, from those really big frog, they just bin cook em up leg part, in a different Country they do it that way. But that *nalja?* How do you cook it?

Lariyari: Cookem with small stones in a coolamon, cookem. When you put em salt now, oh, proper. You can't give it anybody [you don't want to share it].

Maliwa: You naughty girl havem frog.

Lariyari: Good one. And *walaku ngurakin yabayaba* [dingo pups]. Im bin taste like porcupine.

Maliwa: Sister, you naughty girl, havem frog, no good.

Debbie: You still worrying about for that frog?

Lariyari: Mefellow bin always tuck out [eat plenty]. When mefellow bin hunting la bush you know, walkabout.

We never bin going hungry like this time where we sit down hungry fellow. [Like today] Where we wait about for money, that's all. That way we bin go now, wait about for money, that's all, this time. Wait la money bela this now, nothing no more, we no more bin havem *tarnku* properly [the feeling of being full of food], nothing.[8]

8   Kitty Lariyari, tapes 85–86, recorded at Yarralin, 24 July 1986.

Hobbles Danaiyarri

Figure 3.6. Hobbles Danaiyarri, spinning hair string, Yarralin, c. 1985.
Source: Photograph by Darrell Lewis.

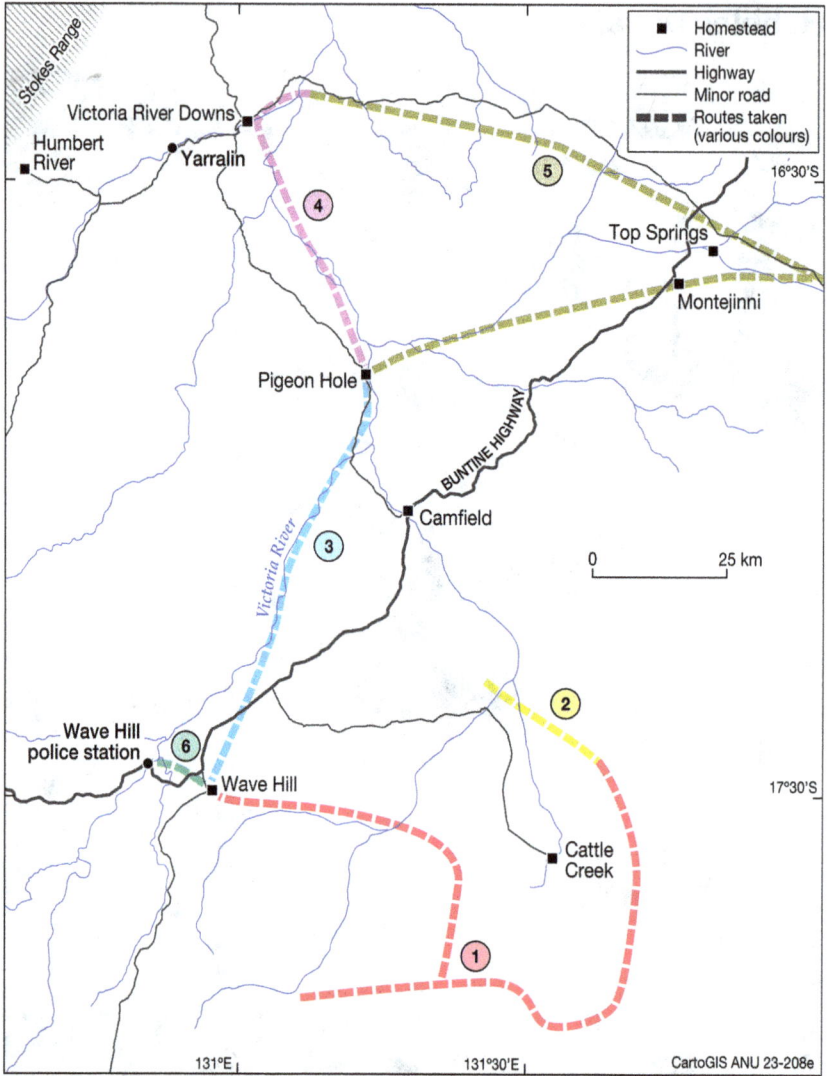

Map 3.4. Hobbles Danaiyarri's footwalk map.

Source: Karina Pelling of CartoGIS ANU.

1.	*Jiya* (kangaroo), *junkuwuru* ('porcupine', echidna), *kalngi* (like a tomato), *kinjirrka* (like a tree seed, root part too), *kitawa* (goanna), *lukarara* (grass seeds, mate for *mangurlu*), *mangurlu* (seeds), *mulyukuna* (black-headed python), *namawurru* (tree sugarbag), *ngurnungurnu* (stone Country tobacco), *pikurta* (bush potato), *walaku* (dingo), *wayita* (small tuber), *yarkajiri* (similar to *kalngi*, like a tomato).
2.	*Wayita* (small tuber).
3.	*Jikamuru* (water lily), *jiya* (kangaroo), *junkuwuru* ('porcupine', echidna), *kitawa* (goanna), *namawurru* (tree sugarbag), *yarkalayin* (hairy yam).

4.	*Jiya* (kangaroo), *kamara* (black-soil long yam), *kilipi* (bush banana), *kitawa* (goanna), *namawurru* (tree sugarbag), *wayita* (small tuber).
5.	Same as 4, including a lot of 'porcupine' (echidna).
6.	Same foods as 4 and 5 including: *jiya* (kangaroo), *junkuwuru* ('porcupine', echidna), *kamara* (black-soil long yam), *kayalarin* (bush onion but different name), *malangarna* (sugarleaf — insect exudate on snappy gum), *muyin* (black plum), *namawurru* and *nankalin* (sugarbag), *nanjalnga* (sugarleaf bloodwood), *nanka* (small animal), *palatmawu* (bandicoot), *punjari* (fig), *tijirrpan* (piebald kangaroo), *wayita* (small tuber).

Hobbles Danaiyarri was born at Cattle Creek on Wave Hill Station in about 1926. His mother was carrying him when she and his father lived out on one of the great plains of Mudbura Country around Cattle Creek, and he spent childhood time there. Although over the course of his life his travels on foot were extensive, his stories were most of all about his own Country at Cattle Creek. His words convey some of the sense of loss that arises with a particular conjunction of displacement and environmental change. When people belong to and are part of their Country and at the same are separated from it, they may experience a sense of loss that is heightened by the suspicion that if they were to return it could all be so different that they would not know it, or, perhaps even worse, it might not know them.

> I was born, and my mother bin bringem [me] out, my father bin bringem [me] out, too long they bin in the desert all day, they bin walk around really biggest that Country. But my mother bin bringem me and that time from desert. That what my mother bin walk around and my father bin walk around. Right up to Cattle Creek stock camp. Soon as my mother and father bin coming back now [to] Wave Hill Station. You know. Where the people live right at that police station. That's where I bin born, right on the police station, Wave Hill.

> Debbie: Were you living in this part when you were kid?

> Hobbles: Yeah. When I was kid, now, coming back now … when I was kid, I bin running into this area all around. Mother bin take me. Father bin take me.

> I was have a lotta good tucker from my mother. We never findem flour and sugar long time ago. Because we bin living really [bush] tucker. Really,

when im bin growing [me] up longa rown [one's own, or their own] land. That's what now. Well we bin havembad lotta *mangorlu* [grass seed], my mother bin fillem longa coolamon. And start to take that *mangorlu* all around longa that coolamon, takem back where that [camp] place. And findem them flat stones, flat one, and killem im tucker there [grind it]. My mother bin get a water and chuckem out, wetenem out that tucker. My mother bin getem flat stone and start to work my tucker now. My mother bin work now. I can see em my mother used her hands and used that [grind] stone too. And really fine nice that tucker. When im bin la really flat one and running down la little rock hole. And that tucker bin still fill im in there. That tucker bin really full, all right. They bin get bark, round one bark, and [make it] just like a pannikin, and get that *mangorlu* then, really [good] tucker. And im bin make lotta fire, and he not put im in the coals, [but] just [on] the hot ground, you know. My mother bin just put that tucker, put im in. And coverem with a sand, you know.

Hobbles vividly described the 'collector ants' who gather seeds and whose seed stores are then raided by people:

Nother *mangari* [vegetable food]—white ant bring him. All the seeds he bring from grass and stack em up like sugar. Just like paw paw seed but more pretty one. Ants been build em up, women take em and put em in coolamon. Take it away to make bread. Make a hole, empty seeds, keep water handy. Got a light stone, grind em up, make a really nice fine flour. Fill up coolamon. Cook in coals, or you can eat it raw, no worries. Cooked one, you never will give it longa somebody [because it tastes so good]. *Mangorlu* [grass seeds] really sweet, that one.[9]

Hobbles discussed some of the plant foods that are part of the desert Country of his childhood:

---

9   Hobbles Danaiyarri, notebook 17, 70.

Well, that *kalngi* [probably a solanum], he's the good tucker like a tomato. You know, when the mother [and] father was show me that tucker, that's the right tucker too. Strong one tucker. That's what we bin havem from beginning. Beginning im growem lotta tucker. That's what my relations [give it] you know, lotta them people bin give me that tucker. *Kalngi,* [and] what we callem *yakajiri.* That's nother tucker, *yakajiri. Kalngi* and *yakajiri,* twofellow same one tucker, but different fruit twofellow gottem. And this one *lukarara.* Nother tucker. Im mate longa *mangorlu* [grass seed] again. Different, [and] still hunting around for *pikurta* [yam]. He's good tucker.

That water, that thing, really good. You can't [go] hungry at all. That's what bin growem up me. After that, oh my mother all gone. My father all gone. Out of the big area and longa this one now.

Hobbles's family left the big plains of the desert Country and became moderately sedentarised on Wave Hill Station. He described going into Country that was new to him, and learning about new tucker, like the little tuber, *wayita.*

Well you come back again now, you come back, you leave that plains again, you come back longa different soil now, you findem what they call im *wayita* [little tuber]. You leavem plains Country now, you come back findem *wayita.*

That's when we come back longa river Country, findem all the *wayita* now. Still all longa ranges country, you know? Good size. You bin leavem what they call im big plain. You leavem. You into the river, too, you go longa other river, you might findem *wayita* there. Righto.

Like others, he expressed memories of plenty: 'We never run out for flour or nothing. Plenty tucker, plenty tucker.'[10] He cherished the foods that had given him body and life, and he linked the knowledge of his own being to the Country and people who nurtured him. Such knowledge is itself life-sustaining, he suggested.

---

10   Hobbles Danaiyarri, notebook 17, 70.

Footwalk politics

Hobbles held the view that demonstrations of local Aboriginal knowledge would convince white people of the justice of Aboriginal people's statements of continuing ownership of the Country. Like many other Aboriginal people, he contrasted his knowledge and care with that of white people's lack of knowledge and care:

> That old woman should be teach im [white people], learn [teach them] to use that tucker, and showem that tucker. Old *kajirri* [woman] bin findem tucker, you know. Well, this kind the really important for people, what we think, you know. We know, we think now, on white man. But still we know Country side. Still we know tucker side. We not careless [forgetting] longa something.

To belong to a place, and to own the knowledge for the place, is to know how to live there. The same point can be made concerning food and water. As Ngarinman man Daly Pulkara said about food:

> We can get im [food], right up end to end [Country to Country]. And some *kartiya* might [be] walking there, [and] he reckon 'no tucker'. And he's walking on top of the tucker now![11]

Riley Young was equally emphatic:

> Two different Law. Nother Law belong to *ngumpin*, blackfella. You know him been walking on this land for many many years. Him been walking by foot, him been carting up him swag la him shoulder. He didn't worry about tobacco, he didn't worry about tucker, he didn't worry about tea, he didn't worry about any kind of feed. Because he been living by sugarbag, bush yam. That kind of a Law him been living. Goanna, or they used to trap em all them birds, gotem spear, kill him, longa water, or any kind of bird. That's all they eat.[12]

---

11   Daly Pulkara, tape 80, recorded at Lingara, 15 July 1986.
12   Riley Young, tape 42, recording date and location unknown.

Under the regimes of colonisation which have so marked people's lives, hunting is also an assertion of autonomy, as Daly contended: 'I don't care about money, that's your way. I got kangaroo, goanna, fish, long as I sitting on the land.'[13]

Good hunters

Good hunters like Jessie Wirrpa are highly valued members of families and communities. Snowy explained to me:

> *Mularij* we call im, proper good shot. Good shot man. Really good shot for everything.
>
> He could findem anything, or killem anything for eat, you know, really *mularij*. When him findem everything, really good shot. Good shot. Findem anything la bush. And might be one time him gettem kangaroo, one time all that *takarin*. That one they call him *mularij* now.
>
> We used to see em some fellow can't hunting kangaroo, anything, well they call him *kuwajal*. That mean he's [just] good enough for get a feed from somebody else.
>
> *Mularij* and *kuwajal,* twofellow.
>
> DR: Is that for men only, or for women too?
>
> SK: For woman, everybody. Woman *mularij* too. Some woman *mularij,* some woman *kuwajal.* Well even man, too, same.[14]

I would never underestimate the physical skills of hunting, as I am completely aware of the limitations of my own skills. While acknowledging the skills of patience, aim, stalking, hiding and all the other physical aspects, it is important also to acknowledge the huge amounts of knowledge that go into making a good hunter. Some of this knowledge concerns co-occurrences in the rhythms of life. Allan Young explained that when the cicadas (*nyirri*) sing, the turtles are becoming fat:

---

13   Daly Pulkara, notebook 3, 52.
14   Snowy Kulmilya, tape 90, recorded at Yarralin, 27 July 1986.

> We hearem *nyirri. Nyirri,* when im sing out longa tree where im sit down. We know, must be turtle, might be come out now. Well that was long time, they go get the spear, them old people. Long as they hearem that *nyirri* talk, they findem walking around, that turtle, they killem gottem spear.[15]

Cicadas are not telling hunters that they should go for turtles, any more than animal tracks tell the hunter that he or she should find and kill the animal. Cicadas hang out along the riverbanks in the big trees there. They tune up their sound apparatus in the hot and humid time of year, and they let loose in a chorus that feels like a physical blow. Being who they are in the world, having a form of action that announces itself vividly in long ear-splitting cries that seem to reverberate inside one's skull, cicadas do what they do. The pattern of the world is that they do this and at the same time turtles are getting fat and becoming more active. The further pattern of the world is that people hear this. If they know what is happening, they grab their spears and go hunting.

Hunting techniques rely on detailed knowledge of animals—their characteristic behaviour, their preferred foods, their times of day and their hiding or resting places. Turtles, for example, eat the fruit that falls off the *japawin* (riverside fig trees, *Ficus coronulata*). People can eat this fruit too, and so do birds and fish. When you go fishing in the hot and wet time of year, you might eat some of the fruit yourself, and then you would throw some in the water to attract turtles or fish.

Different people have different techniques. Riley Young described hunting with a spear:

> And we used to killem turtle gotem wire. Where im always come out eatembad, we used to chuckembad that tucker from fig tree. Chuck im and im come, killem gotem wire [spear with wire point], finish. Oh, big mob we used to kill im. And goanna we used to tuck out—*malajaku.* Or sometimes we used to go sneak up and look la bank, you know, slowly, see im crocodile lying down la sun. Putem wire la im la backbone. That kind of a feed we bin living la bush.[16]

---

15   Allan Young, tape 116, recorded by Darrell Lewis at Katherine, 24 August 2000.
16   Riley Young, tape 86, recorded at Yarralin, 24 July 1986.

Figure 3.7. Riley Young Winpilin trimming a sapling for a spear shaft, Lingara, 1981.

Source: Photograph by Darrell Lewis.

Old Jimmy (Manngayarri)[17] described a more detailed method for killing crocodile:

> You climb the tree, you sit down got a wire, you see it come out, you kill it [spear it]. Kill im, and pull im out outside, and give im bit of *karnti* [wood or stick], and let im bite the *karnti*. He bite that. That *warritja*, you give im stick like that, let im bite that one. Well he can't bite you then. Now you kill im. That way.[18]

Dora described the same method for killing crocodiles and added that when she was a kid people used to catch turtle with their bare hands and toss them up the bank for the kids to kill.[19]

Hunting and fishing stories can go on and on. My purpose here is simply to highlight a few techniques and let them reverberate with other techniques discussed elsewhere (for example, fish traps in Chapter 5 and hunting 'porcupine'/echidna in Chapter 6). They will also have to stand for a rich body of knowledge and stories that in its totality comprises most of daily life, at least for the good hunters.

Dingo

The one exception to the unambiguously happy stories of hunting concerns dingoes. Formerly people in this region ate dingo pups. Today they do not, and this is not only because there may be fewer dingoes because pastoralists lay bait for them, but more significantly (I think) because people who have not eaten dingo find the thought repellent.

Kitty Lariyari was one of the few people I interviewed who spoke of eating dingo pups:

> Dingo? Yeah. I been havem. Montejinni country. Little little one, you know, oh, good one.

> Kitty Maliwa [Lariyari's sister]: No more me. I'm myall [uneducated or unsocialised] one [for eating dingo].

---

17 This is Jimmy Manngayarri, who is a different Jimmy from the one previously mentioned in this chapter.

18 Jimmy Manngayarri, tape 109, recorded at Yarralin, 13 August 1991.

19 Dora Jilpngarri, notebook 6, 21.

Lariyari: Tastes like porcupine, eh, *Junkuwuru* ['porcupine'/echidna]. Mmm.

And *walaku ngurakin yabayaba* [dingo pup-pups]. Im bin taste like *junkuwuru*.[20]

Daly Pulkara spoke of dingo pups, and his experience is more characteristic of people today:

> And one time ago we went up to place called Broadarrow Creek, we found a dog there, dingo. And they went up to make a tucker out of him. Old people reckoned, 'That's good tucker, that.' 'What you call him?' '*Ngurakin.*' 'What do you mean he's the good tucker?' 'Good tucker we eat him.' 'I never tried him. He might make me vomit.' 'No, good tucker that one dingo. [As] Long as you throw the guts away and leave it rubbish.' 'But you eat that?' 'He's good,' they reckon. He might be good, but me, I don't like it head part. That's what I been reckon. And them old woman and my father used to eat that dog, dingo from bush. Yeah, they been havem, gotem big one, but I like that pup-pup.
>
> DR: Pup-pup more better?
>
> DP: Yeah, pup-pup more better. When they roast him, oh, good. I been look it look good when they been roast em. 'Might be good,' I nearly eat him for a while. Oh, leave it till proper really cooked, I tell you. You look good. He might be good tucker too. And my father tell me, 'Yes, they eating that tucker.' 'Yeah?' They used to lost [stop doing] that now. Too much they been eating la *kartiya* [white men's tucker] mostly. Must be. I been used to get along *kartiya* [tucker] too. Plenty longa *kartiya* [white men], I been get all the good tucker, and … look like, 'I don't want to eat any more puppy.'[21]

Snowy's experience was similar to that of his brother Daly:

---

20   Kitty Lariyari, tape 85, recorded at Yarralin, 24 July 1986.
21   Daly Pulkara, tape 80, recorded at Lingara, 15 July 1986.

They been havem dingo but I been never havem. I know … *Ngurakin* they been havem before, them old old women and old old men. Long time [ago]. They been still eat him when I been kid like one of these fellows here. But I didn't like to eat him. We been living on kangaroo. Just that smell, you know, when you get him raw one, cut him up, smell, when you cook him, him like pussy cat again.

DR: Dingo?

SK: Yeah. But we no more like to eatem, you know, because we been growem up some dog there, some small ones. We been think about that one. No more them old women, they been always just cut em up and cookem up and eat him.

DR: When your old people been eatem that dingo, they been growem up first time and then kill him after?

SK: No. That one they been get him that wild one. Wild pup, you know. Kill him whole lot, just cookem. Sometimes they been always get a spear and spear big one. Killem, [and] just put him la fire [to cook].[22]

According to my teachers, only one animal is fully commensal with humans, and that is the dingo. Everything humans eat, dingoes eat too.[23]

Memories

In Table 3.1, I include the numbers of foods that each speaker mentioned. In my view, it is impossible to draw strong conclusions from these figures, as they represent individual variation along with all the other factors. Some people like Big Mick and Dora Jilpngarri loved to talk extensively and exhaustively about detailed knowledge. Others preferred to lump species together and gloss over the detail. Having said that, there may be significance to the fact that the most detailed information was provided by Big Mick and Snowy Kulmilya, both of whom had spent extensive periods of time living in the bush.

22 Snowy Kulmilya, tape 90, recorded at Yarralin, 27 July 1986.
23 Debbie intended to add a discussion of domestication here and to reference Annette Hamilton's work—eds.

A sense of the extensive knowledge of foods can be gained by examining just one list, the foods named by Snowy Kulmilya (Table 3.1).

Table 3.1. Foods remembered being gathered by Snowy Kulmilya.

1.	Namawurru (sugarbag)		31.	Mijat (unidentified)
2.	Jikamuru (water lily)		32.	Karil (bush cucumber)
3.	Wayita (small tuber)		33.	Kitpan (bitter cucumber)
4.	Yarkalayin (water plant with corms)		34.	Malangarna (sugarleaf from snappy gum)
5.	Wajipat (bush potato)		35.	Purrungurn (sugarleaf from red gum)
6.	Ngaruyu (honey)		36.	Lamawut (witchetty grub)
7.	Nankalin (ground sugarbag)		37.	Wak wak (crow)
8.	Yaramalku (tree sugarbag)		38.	Kumira (budgerigar)
9.	Kitawa/malajaku (sand goanna)		39.	Wajilarn (galah)
10.	Warritja (freshwater crocodile)		40.	Garawa (eaglehawk)
11.	Japungarna (water goanna)		41.	Karrang karrang (diver duck)
12.	Kayalarin (bush onion)		42.	Jibilyugu (duck)
13.	Ngamanpurru (conkerberry)		43.	Marpalangpalang (pigeon)
14.	Pulkal (black plum)		44.	Paragnarar (white cockatoo)
15.	Kilipi (bush banana)		45.	Tirrak (black cockatoo)
16.	Kalijpa (pulki, root part)		46.	Wititpuru (spinifex pigeon)
17.	Partiki (nutwood tree)		47.	Jawintingarna (unidentified)
18.	Kuwalampala (turtle)		48.	Kunamulun / kunitjari (olive python)
19.	Yawu (fish)		49.	Jimaruk (water python)
20.	Jiya (rock kangaroo)		50.	Walujapi (black-headed python)
21.	Junkuwuru ('porcupine', echidna)		51.	Walamunpa (pussy cat) (maran = wild)
22.	Kitikiting (like a water goanna)		52.	Jangana (brush-tail possum)
23.	Jamut (turkey)		53.	Jarkulaji (possum)
24.	Warrpa (flying fox)		54.	Palatmawu (bandicoot)
25.	Takirin (like a potato, in sand Country)		55.	Puwun (marsupial mouse)
26.	Korayijkorayij (bush plum or bush orange)		56.	Kakawuli (long yam)
27.	Lamparlampara (wild tea)		57.	Pikurta (yam)
28.	Tilji (wild tea)		58.	Mintarayij (water lily)
29.	Yiparatur (emu)		59.	Walaku (dingo)
30.	Jarrwana (unidentified)			

Source: Author's summary, from recollections of Snowy Kulmilya.

Along with knowledge of foods there is the knowledge of how to cook them. My notebooks are filled with recipes for how to cook foods, and while many of these are urgent, such as the correct methods for leaching toxins from foods, others are specifically to improve flavour. Many of the recipes have Dreaming origins, and many of them are locality specific. For example, the way to cook a turtle, for Yarralin people, differs from the way neighbouring peoples cook the same species of turtle. Differences at this level are constitutive of identity—of home, localised nurturance and belonging.

In some ways the fact that so much of this study is memory work was disheartening to my teachers. They had been nurtured—fed and taught—by their older people, and yet they encountered radically diminishing contexts for feeding and teaching their younger people. Memory, in Hobbles's view, is a form of connectivity between generations of people and Country:

> We remember now, you know. Too many [much] tucker now, white man tucker—biscuit, cold drink, anything, they coverem up [they overlay other knowledge]. But still im remember, still we know … Grandmother and father bin give us beef and tucker, and really making me grow. Give us the tucker, we bin always eat em. Till we bin know. Well, that word now. Father bin do all the best longa teach we proper way. How we put the spear, how we cut the tree for biggest spear, like that. How we going to get the bamboo. We get em bamboo, we know. Ah, here they got a bamboo. Cut em off. Take em up. We get a kangaroo tail, and get a string [tail tendon], tie it on [the spear to hold the point in place]. That kind of thing, we bin know. We remember that tucker too. That word now.[24]

---

24 Hobbles Danaiyarri, tape 91, recorded at Pigeon Hole, 27 July 1986.

Walking with crocodiles

Figure 3.8. Nina Humbert, Lingara, 1982.
Source: Photograph by Darrell Lewis.

Figure 3.9. Nina Humbert's painting of the Jirrikit and Warritja Dreamings (acrylic on canvas), 1991. In the author's private collection.

Source: Photograph by Darrell Lewis.

This painting constitutes another form of memory, situating and sustaining knowledge through visual designs. It is the work of Nina Humbert, Jessie Wirrpa's younger sister, and is the first painting produced in Yarralin using acrylic on canvas. Nina combined the naturalistic style of rock art in the Victoria River region and conventional design elements such as circles for waterholes with a style of dot painting that was brought north out of the desert via Yuendemu, Lajamanu and Daguragu. Jirrikit is the Australian owlet nightjar (*Aegotheles cristatus*), one of Nina's main Dreamings. Here he is shown in the form of a man. He was interested in killing the crocodile that lived in a waterhole in the Wickham River, and so he travelled north into a neighbouring Country in the Stokes Range to get a particular spear shaft made from a shrub that grows in this zone of higher rainfall (*Grewia breviflora*). In the painting we see him with his spears, and we see the crocodile in the waterhole. In the upper centre is another of the Dreamings for Nina's Country: Jimaruk (water snake, probably *Enhydris polylepis*). Her eggs are positioned on either side of her. Along both sides of the painting are billabongs and other ephemeral waters that are connected with the main Jimaruk billabong, and with the main waterhole on the Wickham River. Jimaruk was walking back and forth among the waterholes. She made a camp (a site now known as Jimarukala) and put her eggs there. If you 'bust' the eggs, snakes will come out everywhere. Jimaruk was looking for permanent water; she went up to two billabongs but couldn't make a big hole, so she went back to the river and there she changed over into a Rainbow Snake. She continues to live in the permanent waterhole.

The crocodile here is the freshwater species (*Crocodylus johnstoni*). It differs from the larger saltwater crocodile (*C. porosus*) in several ways that are significant to humans, the most important of which is that it does not hunt people. Its bite can be dangerous, but Victoria River people swim with freshwater crocodiles all their lives without mishap. By contrast, saltwater crocs are predators for large mammals, taking people regularly, although far more tourists than locals are taken.

While by no means exhaustive, this painting tells a lot about a portion of Nina's Country and Dreamings. You see the main waterholes and their connections to the more ephemeral waters, and perhaps you glimpse the importance of knowing about permanent, ephemeral and subterranean waters. Jirrikit with his spears shows a trade relationship with a neighbouring Country. The stories themselves make more connections outside the painting. This is to say that, like the tin cans at the Nanganarri billabong, the stories from this place work their way outward into the world: the crocodile was killed and was carried up into the sky where he can be seen today in a set of stars; the boys who killed the crocodile are still in the area as a stone. The story tells of moving waters. When the river goes down and the permanent waterhole near the homestead at VRD is shallow, you can see the two boys and Jirrikit's whiskers there in the bed of the river. When the river comes up the Dreamings can no longer be seen, but when the river goes down again, there they still are.

Nina's painting also refers indirectly to social history: Jirrikit was there in the area in the form of a rock, but when they built the airstrip at Victoria River Downs station they knocked him over (Rose 1992, 108–9). Her painting shows us eco-place in its Dreaming presence, and in its connectedness to other places. As I will discuss in later chapters, this work speaks as well to seasons and communication. It thus shows us connections in the ephemeral world.

It is important to think about how much we cannot see. Most of what I have described was explained to me by Nina and Jessie. The knowledge that people bring to their Country fills in the ambiguities that are held present in the designs (see also Lewis and Rose 1988). As Eric Michaels (1993) and others have shown, Aboriginal art privileges the living, and upholds the rights of knowledge held by senior people. Another frame for this painting is the fact that it was painted by Nina: she painted her Country and Dreamings because these are the places, Dreamings, stories and knowledge of events that she has the right to condense, depict and share with others.

Yet another part of Nina's painting is its implicit references to seasons (Chapter 6). When the two kids tormented Jirrikit, he pulled a bit of his whisker, and sang up a big cold wind. That wind scoured the trees off a big black soil plain. The two kids saw the wind, and ran for their lives, but the wind caught them and carried them up into the sky, and then threw them back down, dead. It took the crocodile up into the sky, and there it remains today.[25]

That constellation is probably the Southern Cross. You do not actually see the crocodile; you just see men standing around it trying to spear it. When the constellation jumps up in the sky it is cold time. The big winds and the cold weather come together with Warritja, the crocodile, the constellation.

---

25   Big Mick Kangkinang, notebook 39, 79–82 (as told by him).

Now, look again at Nina's painting and consider that it is figuring motion. From the perspective of motion, Jirrikit is travelling, the crocodile is swimming, the boys (who are not shown) are plotting to kill it, the water snake is laying her eggs, the waters are flowing back and forth. It thus spills out into time—the eggs will hatch; the rain will fall to replenish the ephemeral waters. And the Country is being cared for because here is the evidence of an owner who knows her Country. The painting shows time, place, motion, care and renewal, condensed on canvas but actualising in the real world of eco-place.

Looking at the painting as a condensation of motion implicates the viewer as a participant. If this is a glimpse of the happening world, how does the viewer fit in? The painting can thus be seen as an invitation to encounter the people who will take you to the places and teach you the stories, perhaps even show you 'something'. While preserving the authority of people whose responsibilities are to these Dreamings, these places and this knowledge, the painting is also a call—an invitation to become present.

A footwalk perspective suggests that structure is best understood as pattern. Structure can be abstracted, pattern is embedded. My shift from structure to pattern signals a movement away from concepts of stasis and reproduction of stasis, in favour of concepts of recursive motion. In this and subsequent chapters I will be proposing that pattern, while predictable, is also ephemeral. An analogy is with ripples caused by a stone in a billabong. The ripples are predictable, but their occurrence is contingent. In order for patterns, like ripples, to remain, contingency must be charged up through the recursions of life in the changing and ephemeral world. The stones must be thrown.

4

Dreaming Organisation

Before *kartiya* [whitefellas], blackfellas bin just walking
around organising the Country.

Hobbles Danaiyarri

Creation

People's work to keep the Country 'organised' builds on a foundation of
Dreaming tracks and connections. Dreaming geography is the creation
of patterns that both differentiate and connect. Each Country contains
a plurality of sites, and the sites are connected by tracks; the ceremonies
and ecologies that are part of the tracks work across bounded Countries.
Creation elaborates the tracks, and thus elaborates the intersecting and
crosscutting patterns of connection between eco-places.

Each enclosed area is a notional Country.[1] Dots are Dreaming sites; arrows
mark the direction of Dreaming travel. Some of the dots are connected
by Dreaming tracks, which show a system of crosscutting connections
(between Countries and sites). The significant features are: no Country is
unconnected to others; and no Country is connected to all others. Each can
trace a connection to another along a Dreaming track that is crosscut by
another track. Thus: A is connected to B, C, D and J along one track, and

---

1   While Debbie references a diagram to show how these connections work, the editors have not been
able to find this figure in Debbie's archive. They have looked over published papers and can't it find
there either.

to H, E, F and G along another track. A has no direct connection with I. However, I is connected to a number of other Countries to which A is also connected: G, C, H, E, J, D, C and B.

The significance of this pattern of differentiation and connection is far reaching. It constitutes the organisation of geography and seasons and is discussed in the next two chapters where the focus is on precision (boundaries/differentiation) and on patchiness (crosscutting and recurrent bits). It constitutes a patterned ground of ethics (Chapter 6), and, I will suggest, constitutes a sustaining pattern for time and thus for serious life (Chapter 5).

In a real-life set of Countries and tracks, there would be complexity: contiguous Countries would be connected by the short tracks of local Dreamings, and on a regional basis there could be one Country into which many tracks converge. Experientially, walking in Country means walking in ecosystems.

There is also a greater density of connections among contiguous Countries. Dreamings travel, and their essences remain in Country. Thus, each Country is connected by Dreaming to other Countries in patterns of cross-penetration. With localised Dreamings, Country A Dreaming travels into Country B and returns, so that a part of Country A is in Country B. Similar relationships obtain between A and C, etc., etc. The same process obtains from the perspective of Country B. One or more of its Dreamings travel to Country A and return, so that a part of B is in A. Patterns repeat across Countries, so while there is no controlling instance of connection, the patterned repetitions become a system that both resists domination and resists disorder.

David Turner discusses a pattern that is pervasive in Groote Eylandt geography and personhood—a portion of each is located in the other without loss of integrity to either (1996, 14).[2]

The discussion of pattern is applicable to landforms and plant and animal communities as well. They are not there as random events, but rather emerge from and form organised patterns. I will look first to the boundaries between three main ecological zones, and then I will examine the crosscutting and overlapping distributions that bring both patchiness and

---

2   Debbie had a note here indicating that she wanted to add a paragraph on Turner's work and to link it to patchiness and precision—eds.

connectivity into a system of organised difference. Dreaming organisation of life forms and Dreaming demarcations of difference and connection underlie ecological zones.

In the Victoria River Country of my teachers, there are three big ecological zones: the 'saltwater side' zone, marked by the tidal influence in the big rivers; the big freshwater Country of the inland (*laman*), where the rivers are fed by smaller creeks (*pinka*) and finally flow into the sea; and savanna desert (*kaja*) which is marked by an absence of permanent surface water. Following the river inland, each zone is adjacent to yet another zone: on the saltwater side there is an adjacent coastal zone comprising floodplains, mudflats and swamps. On the savanna desert side there is an adjacent zone of more arid desert marked by sandy plains and sand dunes (sandhills), and by rivers (if any) that flow inland, fanning into floodouts and disappearing into the ground.

Each of the zones in this study is uniquely defined by water: salt or fresh, present or absent, direction of flow. Furthermore, each has its unique plants and animals, and each can be defined by one or more indicator species. In this region, these broad zones extend west and east for much greater distances than they extend south and north. They thus correlate with rainfall patterns and with proximity to the humid and wetter coastal zone (to the north) or the arid inland (to the south).[3]

My perspective in this analysis is situated in the inland riverine Country where the greatest amount of my time has been spent. My teachers included people from all three zones, but all my great teachers were themselves situated in riverine Country during the time I spent with them, except when we travelled to other places. Thus, while those who came from elsewhere were well able, and often very interested in, teaching me about their home places, their perspectives also were situated in the riverine Country where we spent most of our time together. The effect of having this perspective is that I learned far more about how the saltwater side and savanna desert Country are marked as different from the freshwater riverine Country than I did about how neighbouring people define the freshwater Country.

---

3   From an outsider's perspective, they form bands that extend across river systems and state boundaries, and thus across both natural and political boundaries. From the perspective of Aboriginal people, their own zones extend within a major catchment area.

Saltwater and freshwater

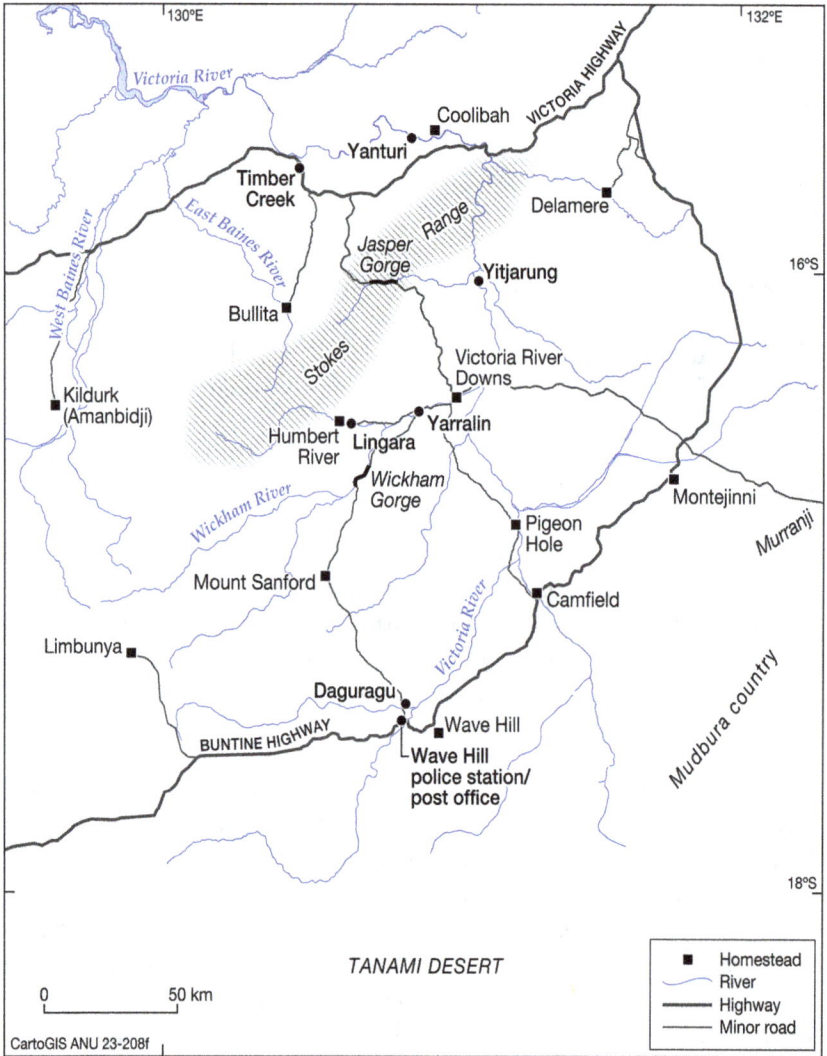

Map 4.1. The Victoria River region showing the main ecological zones as defined by Aboriginal people—the 'saltwater side' north of the Stokes Range, the 'river Country' (*laman*—river and *pinka*—creek), below the Stokes Range, and the savanna desert (*kaja*), south of the river Country.

Source: Karina Pelling of CartoGIS ANU.

Figure 4.1. Yanturi on the Victoria River, a Dreaming place for Barramundi, Plover, Pigeon and others, Coolibah Station, 1982.

Source: Photograph by Darrell Lewis.

From the riverine perspective, there is a well-defined boundary to the north, and the northern area is called 'saltwater Country' or 'saltwater side'. The Dreaming Pigeons followed the Victoria River north out of the desert fringe carrying large slabs of sandstone. The pieces of stone that they dropped formed the great sandstone cliffs and outliers along the Victoria River. These seed-eating birds used the sandstone as grindstones, and they were demarcating a zone of seed-grinding technology. The last place where they dropped sandstone is at a site in the river called Yanturi. Here there is a large drop in the riverbed; in the early dry season the waterfall is magnificent. This is the boundary for salt water: tidal influence extends inland within about 20 kilometres of Yanturi. This is also the approximate northern extent of grindstone technology. From here northward, mortar and pestle technology predominates. I was told this version of the story by Old Jimmy, and he took a decidedly inland view of the pigeons' activity and focused in particular on the saltwater crocodile, locally known as the alligator. In Jimmy's view, the pigeons put Yanturi there as a blockade 'to keep out alligator and keep out all those cheeky saltwater things'.

Yanturi marks the tidal influence, but it is not an absolute boundary. When the Victoria River is up, it covers Yanturi, and fish as well as other 'cheeky saltwater things' swim upstream. Further inland, the barramundi (giant perch, *Lates calcarifer*) marks another boundary. This species can live in freshwater for much of the year but migrates to the estuaries to spawn (Larson and Martin 1990, 42–43). The Barramundi Dreaming travelled upstream from Timber Creek, trying to pull the salt water through. It was unable to pull the water past Yanturi, but the Barramundi itself travelled upstream as far as the Wickham Gorge and other inland sites. Its travels mark the extent of barramundi migrations today and connect inland freshwater systems with the estuarine systems closer to the coast.

Up there at Yanturi, the Country belongs to Ngaliwurru and Nungali people. Their story is a bit different and is focused on several aspects of the place. One of the main features is the large rock hole at the base of the rock wall. This rock hole was made by the Dreaming Plover, and there is a Dreaming tree (*Eucalyptus camaldulensis*) that is the Barramundi who tried to jump over the cliff to meet up with the other Barramundi who had swum inland. The rock hole is also a rich fishing site: people spear or catch stingray, sawfish, shark, alligator (crocodile), barramundi and other mainly saltwater species, along with all the freshwater species that move around here. Yanturi was a site for ceremonial gatherings in the old days. The richness of the food supply here just after the rainy time enabled people from all the neighbouring groups to gather: Nungali, Ngaliwurru, Jaminjung, Wardaman, Karangpurru, Wulayi Ngarinman and Wickham River Ngarinman peoples.

Another boundary between the big river Country and the saltwater side is formed by the Stokes Range. The clarity of this boundary derives both from its physical presence and from the way it blocks some of the rain coming in on the monsoon. The range lies south of Yanturi between Timber Creek and Yarralin. Rainfall figures are compiled at Timber Creek (north of the range) and at Victoria River Downs (VRD) station (20 kilometres east of Yarralin, and south of the range). These figures show a difference of about 235 mm of annual mean rainfall: 632.6 mm for VRD, compared to 867.7 mm annual mean rainfall for Timber Creek.[4] To these aggregated figures must be added the fact that VRD gets slightly more rainfall in the dry season than does Timber Creek (6.1 mm during June, July and August, compared with

---

4   These figures are contemporaneous at time of writing, c. 2003—eds.

3.4 mm at Timber Creek). This means that at VRD slightly more rain falls at a time of year when it is less likely to promote the growth of plants and the sustenance of animals; the inland river Country is thus drier than the annual mean figures would suggest.

Along the Stokes Range two indicator species are especially significant: the boab (*jamulang*), and the long yam known as *kakawuli*. The boab (*Adansonia gregorii*) reaches its current southern limit on the south side of the range. *Kakawuli* (*Dioscorea transversa*) occurs north of the range and part way into it, but not on the south side. Dora discussed crossing this boundary when she entered a region where the staple food was unknown to her (Chapter 3). I have already mentioned the spears made of *Grewia breviflora* that Jirrikit went searching for in the Stokes Range. The Stokes Range seems also to have been the southernmost range for gliding possums (*Petaurus breviceps*), a species now apparently extinct locally.

As well as constituting a zone of ecological differentiation, the range marks a major linguistic boundary, dividing two different families of languages. It is also a cultural boundary, as it marks some differences in kinship, marriage and other cultural patterns (discussed in Rose 1992, 118–19, 1998). Social and cultural boundaries were no obstacle to social interaction, however. People from both sides of this social boundary gathered at Yanturi.

Another gathering place is located just south of the range at a site called Yitjarung. As at Yanturi, people from neighbouring groups gathered, moving across ecological, social and cultural discontinuities in order to interact. Yitjarung is in Jessie Wirrpa's Country. Her family travelled there from VRD following the track of their Dreaming ancestor Jirrikit. He was going north into the ranges to get spears that were flexible enough not to shatter on impact with a freshwater crocodile, and he had to get them from people on the other side of the range. When Jessie's people travelled to Yitjarung they met up with people from the ranges Country, Big Mick's family and others, who were the suppliers of the spears Jirrikit had sought. Jessie described these intertribal gatherings:

> Jessie: Jasper Gorge and VRD river. We met up there now. That Country belongs to us, my *kaku* [father's father] Country. That's for us, as far as that now. We used to be meeting everybody.
>
> Big Mick: Kuwang mob. Yeah, Kuwang [Ngaliwurru —ranges Country].

Jessie: Ah, Karangpurru mob, *kaku* mob, we were camping out there. And after, everybody went back home. And we come back this way [to VRD]. And *Wardaman* [people] go back home. They had the holiday at the river. Yirtjarung, meet up there.[5]

The saltwater side is relatively well watered and given that Big Mick's home Country is in the Stokes Range, I asked him once to tell me all the kinds of water. His list included these types: spring water, flood water, running water, round billabong, *laman*/river, little creek, junction, soak water, rock hole, ice water, fog.[6] Many people spoke of this factor when they discussed the colour of the grass. On the north side of the range the grass is green during much of the year, but on the south side it is yellow. You drive through Jasper Gorge, cross the creek, and come out onto the flats on the other (eastern) side of the gorge, and the country opens out into yellows, golds and bronzes, along with the silvery greens of spinifex and the dusky greens of the eucalyptus leaves.

Freshwater riverine and savanna — *kaja*

The ecological and social boundaries between riverine and savanna desert are not as clearly defined as those between saltwater and freshwater. There are no major geomorphologic barriers such as the Stokes Range, and rainfall figures also tell a story of greater continuity. The difference between VRD and Timber Creek is significantly greater than the difference between VRD and the Wave Hill Post Office, a few kilometres from Daguragu. VRD's 632.6 mm annual mean is only 165 mm greater than the 467 mm annual mean rainfall for Wave Hill/Daguragu. Wave Hill gets slightly more rain than VRD in the dry season (8.7 mm in June, July and August, compared with 6.1 mm at VRD), and is correspondingly a bit drier overall.

Kaja is glossed as desert and is defined as an absence of water. Daguragu is in riverine Country because of the Victoria River, but to the south, east and west of Daguragu the Country is *kaja*. The *kaja* is still savanna, with some large treeless grass plains. It is differentiated from the more inland desert by its landforms, savanna vegetation and animal species, as well as by linguistic and other factors. My study includes savanna *kaja* but not

---

5   Jessie Wirrpa and Big Mick Kangkinang, tape 78, recorded at Yarralin, 13 July 1986.
6   Big Mick Kangkinang, notebook 12, 23.

sandhill *kaja* (Tanami Desert, for example). Within the savanna *kaja* zone the different languages are members of the same language family, and the people hold kinship, marriage and ceremonies in common (discussed in Rose 1992, 118). The distinction between savanna and sandhill desert peoples is marked by linguistic and other social and cultural differences, as well as by numerous floral and faunal differences. The main body of my research took me into Mudbura Country to look at differentiation, and so I will focus on that area.

Mudbura *kaja* — savanna (desert)

The Australian continent is bisected from north to south by a gently rising plateau. In the northern section, that concerns us here, this dry plateau country divides catchment areas. On the east side the rivers flow into the Arafura Sea and the Gulf of Carpentaria. On the west side they flow into the Timor Sea and the Bonaparte Gulf. Mudbura Country straddles the western flank of the plateau in Country where there are no rivers. The southern part of Mudbura Country is full desert. A band across the middle is home to the grassy treeless plains that ensure that Mudbura Country is rich with tucker. When the rains do fall, the water has nowhere to go. The waterholes fill up and overflow into the plains, creating huge wetlands. Waterbirds flock to Mudbura Country to nest, and the plenitude of birds, eggs, fruits and other tucker is awesome. The great lakes such as Lake Woods near Newcastle Waters, dry for much of the time, become home to pelicans, brolgas, jabirus and a myriad of other large waterbirds as well as flocks of ducks and other smaller birds. Many of the Dreamings in this desert Country are waterbirds.

The segment of Mudbura Country in which I have carried out ethnobotanical research is the northern segment. The botanist Jeremy Russell-Smith describes this region as an arid forest. It is well wooded with contiguous but separate stands of lancewood (*Acacia shirleyi*), bullwaddy, and in the limestone areas snappy gum with a beautiful understorey of yellow, prickly, aromatic spinifex (probably *Triodia pungens*). The famed Murranji droving track cuts through this arid forest. It was designed to move cattle in long hauls from one waterhole to another. They wrecked the largely ephemeral waterholes in due course, and the government installed bores. In 1990, Darrell Lewis and I did a quick survey across the old track; Darrell documenting historic sites, and I seeking to document some of the major aspects of Mudbura ethnobotany, particularly as they pertained to questions that had arisen in my VRD research.

My main teachers in this portion of the research were Nugget Collins Ngurrartarlu and Long Captain Marrjala. One of the questions at the forefront of my mind was how people managed to travel through the densely dangerous bullwaddy scrub, and how they managed to subsist without large permanent water sources. From a riverine perspective, as Old Jimmy kept emphasising, *kaja* Country has no water. From the perspective of the people who belong there, there is water all right, you just have to know where it is and how to access it.

Formerly, Nugget explained, before white settlers pushed their way through the Murranji Country, the arid forest was crisscrossed with walking paths. People followed tracks to their destinations, and the paths articulated with waterholes and soaks. The Murranji waterhole is the main one, and there are several others. Where there was not surface water, people had dug wells. Much of this part of the country is limestone. Water accumulates in this underground country, and people tapped into it with their wells.

Long Captain's Country was south of the arid forest, in even drier country, with no permanent waterholes, on his account. He had not footwalked his own Country, as his family had become caught up in cattle station life, but when he used to go across the Murranji droving, his relations would teach him about his Country to the south, telling him of soaks where people dug for water. They also described a tree called *karrinbirri*. These trees were hollow and collected water in their trunks during big rains. At night, they said, you could see something like lightning hovering around the top of the tree. You would mark that tree, and in the morning come back and make a small hole near the base of the tree to allow the water to run out into your container. When enough water had been collected or all the water was gone, the hole would be plugged with a stick so that the tree would hold water again during the next big rains (see Lewis 2007, 9–10).

From the riverine perspective, one indicator species for desert is the burrowing frog (also known as *kajangarna*, meaning 'desert dweller'). Many of the riverine people differentiate themselves by refusing to eat frogs, as Maliwa demonstrated when she scolded her sister for eating them (Chapter 3). Old Jimmy associated frogs with Cattle Creek country, and thus used them to distinguish the Mudbura *kaja* from his own Country, which is also *kaja* in some parts.

> All around Cattle Creek … they lived on the *kayaman*.
> *Kayaman*, it's a little *ngalpung* [frog]. *Kayaman*.
> They call it little *kayaman*. Little frog. That's the one

they eat, early days, you know. I never ate it. I don't know that one. They all like that, though, they get hundreds in the one hole.

A second indicator species for savanna *kaja* is *miyaka* (*Brachychiton* sp.). The edible seeds of this tree are an important food source, not only because of their abundance, but also because of timing. They become ripe in a time when relatively few other vegetable tuckers are available. Although the *Brachychiton* species are widespread, *miyaka* only comes into abundance in the savanna desert, and in the Victoria River Country it is particularly associated with the desert Country on the eastern side, that is, Mudbura Country.

Charcoal Winpara's Country is in the Mudbura *kaja*. He described an area in which the plant grows, but then went on to identify it as his (desert) Country: '*Miyaka,* that's for desert country. In desert country. Ah, plenty down there, that's my Country, from Montejinnie, nother side. You see plenty *miyaka* there all round.'[7]

These boundary markers are experienced. As you move through Country you see, and eat, different plants and animals. Coming from savanna *kaja* into freshwater riverine Country, Hobbles encountered *wayita:* the little tubers that were left there by the Nanganarri Women. Hobbles associated it with soil type. According to Nicholas Smith and colleagues (1993, 47), they are always found growing in the black soil of the river Country.

Similarly, Nugget pointed out a highly visible marker when one travels from the Murranji savanna desert into the river Country. The plateau grades downward so gently that you don't notice it in a truck. Then you come to a major decline—the 'jump-up'—only travelling west you actually 'jump' down. Almost immediately you start to see the inland bloodwood (*Corymbia terminalis*). It is not the case that this species is confined to the Country west of the jump-up; it is widely distributed across much of inland Australia. However, it is rare or non-existent in Mudbura *kaja* Country. Experientially, it tells you that you are into the riverine Country.

In contrast to the distinctions between salt and fresh water and between river and savanna *kaja,* the sandhill desert is distinguished by a large number of different species. The Dreaming Wallaby, for example, travelled south into the desert. When he got into the sandhill Country, he changed over into

---

7   Charcoal Winpara, tape 84, recorded at Yarralin, 21 July 1986.

the Big Red Desert Kangaroo (*Wiwiri*). Other significant markers of drier desert Country include witchetty grubs (discussed below), solanum species and a number of lizard species.

Mosaics

The three big zones—saltwater side, inland riverine and savanna desert—are both bounded and crosscut. On the one hand there are the unambiguous differences. The sweet long yam, *kakawuli,* does not grow south of the Stokes Range and is thus an unambiguous mark of Country to the north, from an inland perspective. Similarly, *miyaka* (*Brachychiton*) is a savanna desert tucker and does not grow in the riverine Country; from a riverine perspective, it unambiguously signals desert.

Zonal differentiations are crosscut by other contrasts and the zones themselves are filled in with fine-grained detail. One type of crosscutting works with elevation at a local level. The contours of the land recapitulate locally some of the major contrasts associated with broad zones. Jessie Wirrpa listed the trees that grow 'outside' or 'top side' (hills or high ground away from the rivers, *kangkula*) in contrast to those which grow 'bottom side' or river flats or low ground close to the rivers (*kunjura*). Her analysis applied to her own freshwater riverine Country. She discussed three major communities: the riverside community includes river red gums, paperbarks, pandanus and other riverine trees, along with pockets of monsoon forest or 'jungle'. While there are of course no saltwater mangroves, freshwater mangroves (*mawunji, Barringtonia acutangula*) are plentiful. Out on the flats there are the characteristic tropical savanna communities of the inland riverine Country: scattered eucalypts and a few other trees including *bauhinia* and nutwood tree, along with grass and shrub understorey. On the dry and stony slopes of the mesas, and on top of the mesas and cliffs where there are no creeks or rivers, one finds plant communities that are characteristic of the savanna desert.

It is clear that the distinguishing features of zones enable a person to know, in their travels, that they have crossed a boundary. Within a given area, however, clusters of members of these different zonal communities are interpolated and thus form landscapes characterised by patches or mosaics. Some plants that are characteristically for savanna *kaja* will be located on top of the mesas, and some varieties of spinifex that are characteristically for *kaja* Country will be located on stony slopes. In light of local interleaving of clusters associated with zones, the significance of unambiguous indicator species that are not located beyond their own zone becomes clear.

Another aspect of the interpenetration of zonal features is that of plants that are widespread, but only start to flourish in a particular zone or a particular Country, and thus become associated with that place. An example is the tuber known as *pikurta* (*Ipomea costata,* Wightman 1994, 34), which from a riverine perspective is characteristically desert tucker. It also can be found in the Wickham Gorge where it is not abundant but is sufficiently established to be known. Other plants are widespread but their appearance changes. For example, *mulurmi* (turpentine wattle, *Acacia lysiphloia*) grows in both the riverine Country and the desert. In the riverine Country it is a tall scrub, and in Mudbura Country it is a bush; in both places it is prized for its medicinal qualities.[8]

Another food that is distinguished by density and prominence is the witchetty grub. Grubs live in a variety of trees and shrubs. In the savanna desert (and throughout the arid zone) they are prominent subsistence foods and are characteristically found in the roots of several acacia shrubs. In the riverine Country they are found in trees. There they are not a major food item, they do not figure prominently in ceremony, nor are they associated with the seasons. In the desert Country, though, as Hobbles explained: 'That bush is in desert Country, and wet time, you can pull it out and find grubs underneath. That is truly *ngarin* [meat]. Wet time really belongs to grubs.' Old Jimmy added to the story:

> And witchetty grub, I'm not talking about this that
> lives on the river, not that witchetty grub that lives
> in the white gum tree [*punpu*]. I'm talking about the
> one that lives in the desert. It lives in the—oh, you
> might see a sort of a little tree. You move that one,
> you pull that one, you'll see ten or twenty witchetty
> grubs there [in the roots].[9]

Another way in which the broad zones are crosscut is by soil types and other geomorphologic features. Among the many soil types that my teachers identified are red soil, black soil, limestone, sandy soil, sandstone and stony ground. These types intersect each other across zones, and many have their characteristic plant communities or species.

---

8   Dora Jilpngarri, notebook 53, 149.
9   Jimmy Manngayarri, tape 114, recorded by Darrell Lewis at Daguragu, 19 August 2000.

Steep hills and gorge tops are the homes or habitats for numerous species which have a counterpart in another habitat. The fig tree of the stony hillsides (rock fig, *Ficus leucotricha* var. *leucotricha*) is a mate for the riverside fig (*Ficus racemosa*), as I will discuss in Chapter 6. For the moment my interest is in relationships across habitats. Old Jimmy explained the idea in reference to bloodwood trees, one for river Country and one for *kaja*. We had seen each of them recently, one on the flats, and one on the high dry Country along the top of a gorge:

> This sort of bloodwood, we call im this one bloodwood, bloodwood this one, river bloodwood [*C. terminalis*]. That one we call im *kajangarna* [desert dwelling], that one *puwaji* [*C. dichromophloia*].[10]

Yet another form of crosscutting is worked out through plants. Most of the mateship between plants identifies relationships based on the interplay of sameness and difference across habitats and zones. Difference is identified and then is crosscut by similarity; similarities are identified and crosscut by differences. Contiguity also seems to be a factor. The result is a mosaic of connectivities within and between genera and moving across habitats, landforms and ecological zones.

A plant like *kakawuli* that belongs to the saltwater side is referred to by the same term in Ngarinman (riverine) language because there are no Ngarinman (riverine) *kakawuli*. However, where species are located across zones and languages, there are some shifts in terminology which produce the mosaic effect of same and different. *Jartpuru* is a good example. In the riverine Ngarinman nomenclature of Yarralin and Lingara, *jartpuru* is a bloodwood tree, known in the English vernacular as the 'inland bloodwood'. This same term applies to this same species in Bilinara and Gurindji languages. In Mudbura Country, however, this tree is either extremely rare or non-existent. A similar bloodwood (*C. dichromophloia*), 'variable barked bloodwood' or 'mountain bloodwood' in English vernacular, is called *jartpuru*. In Ngarinman and Gurindji languages, *C. dichromophloia* is called *puwaji*.

Mateship also shifts across zones. In Ngarinman and Bilinara Country *wanymirra* is mother to *wayita,* but Glenn Wightman (1994, 53) reports that amongst his Gurindji teachers at Daguragu *wanymirra* is the grandmother for *wayita* (*wayij*).

---

10   Jimmy Manngayarri, tape 111, recorded at Yarralin, 14 August 1991.

Dreaming tracks produce another, highly significant, form of crosscutting. The Dreaming track for the uninitiated men, for example, is marked by snappy gum trees that are 'pegged out just like a line'. The trees march from the saltwater side down through the riverine Country and toward the desert, following the landforms and soils that make up their habitat.

Dreaming actions

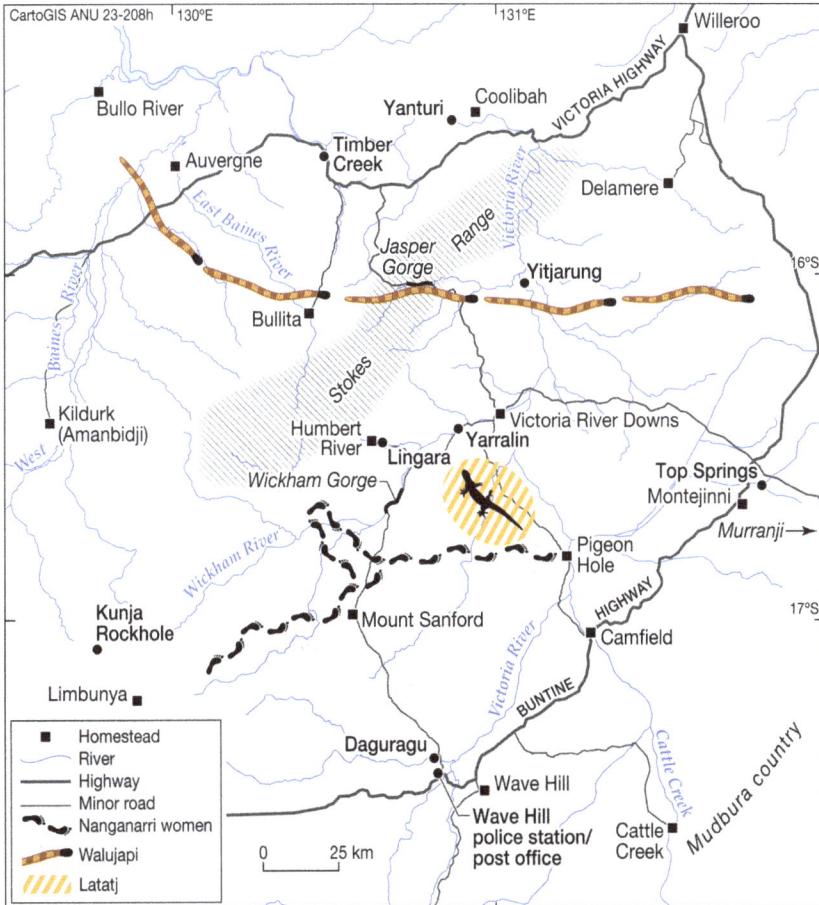

Map 4.2. Dreamings that signify a zone (Latatj), connection between patches (Nanganarri) and demarcate a boundary (Walujapi).

Source: Karina Pelling of CartoGIS ANU.

I will discuss several Dreamings in order to explore the idea that bioregional communities are themselves the marks or evidence of connection, continuity and association. I will examine four processes: negotiation of discontinuity, the creation of a zone (Latatj Dreaming), the connection between patches (Nanganarri Dreaming), and the demarcation of a boundary (Walujapi Dreaming).

Discontinuities

The interpolation of soils, landforms and associated species produces a mosaic effect within and across the major zones. There are, in addition, discontinuities that are said to have been established by Dreaming. One such discontinuity was established by the Sugarbag Dreamings who fought over who could go where. One was a large one and belonged to riverine Country, the other was small and belonged to savanna *kaja*. As Ngarinman (river) people tell the story, their large sugarbag was trying to get into the desert Country, and the smaller desert one hunted it back. The boundary between the two is in the watershed range Country between the headwaters of the Humbert River and *kaja* Country to the south-west. Two of them were fighting, ground sugarbag and tree sugarbag, and where one killed the other and thus pushed it back to the north there is both a social boundary between Bilinara and Ngarinman languages and an ecological boundary between types of sugarbag.[11]

Even the reaches that exceed boundaries are still imagined in the mode of discontinuity. I discussed the Barramundi Dreaming that swam upstream from Yanturi. In Dreaming geography there are now several Barramundi Dreamings—one is at Yanturi where it tries and fails to jump up over the cliff. Others are in the inland waterways where they struggle to return to the salt water but are blocked by the fact that the rivers have shrunk back to disconnected waterholes. The Barramundi Dreaming in Daly's Country around the Wickham Gorge struggles, and fails, to reach its mate at Yanturi.

The maps for such landscapes are complex, indeed. Old Jimmy always held that the maps for these relationships are in the ground itself:

> I mean like, you know, on this Earth. On this ground, isn't it. You know, what the Dreaming has done. This ground, it's sort of a map. Map. Map for the people.

---

11 Jimmy Manngayarri, notebook 55, 117.

That way, this ground here … On this ground—
Janja. Ground. Well that's that map for the *ngumpin*,
right on this ground here. *Janja*—that's the ground
where we sit down.[12]

Latatj

Latatj is a small goanna (Storr's monitor, *Varanus storri*), known locally as a
rock goanna. According to one of the main scientific sources, the Australian
population is divided into two widely separated groups: the eastern group
is located primarily in Queensland with slight overlap into the eastern
Northern Territory, while the western group straddles the Country along
the mid and upper Victoria and Ord Rivers and extends into the inland
Kimberley (Swan 1995, 102). The western range of the *latatj* very neatly
coincides with a portion of the big river and *kaja* Country, stopping short of
the saltwater influence in the north and the sandhill Country to the south.
The range thus includes the region in which the majority of my research has
taken place.

According to Old Jimmy, Latatj was responsible for organising the plant
and other communities that go to make up this region. I asked him once
about a plant I was hoping to be able to identify called *tipil*. I knew that it
was supposed to grow further north, and I wondered if it might also have
grown in some remote corner of his Country. He rejected the idea that
I would find *tipil* in his Country:

> No, nothing. That Latatj didn't want that kind of
> tucker. All kinds of tucker, he pushed them back,
> right back that way, on the sea side. He must have
> a good fish, good crocodile, good sugarbag [native
> honey], everything, good way …
>
> He's boss for every tucker there. Whatever he don't
> want, he push it away: 'You take it back,' he said.
> 'I don't want that one.' He don't want it. Latatj don't
> want it: 'Keep it out on that side, on the sea side.'[13]

---

12 Jimmy Manngayarri, tape 110, recorded at Yarralin, 13–14 August 1991.
13 Jimmy Manngayarri, tape 109, recorded at Yarralin, 13 August 1991.

Figure 4.2. Debbie, Ivy Kulngarri and Nancy Jalayingali at Winingili, a Latatj Dreaming site on the road between Yarralin and Pigeon Hole, 1982.

Source: Photograph by Darrell Lewis.

Figure 4.3. Rock painting of Latatj, the rock goanna (natural ochres), 1984.

Source: Photograph by Darrell Lewis.

On another occasion I asked Jimmy whether there was in his Country the kind of sugarbag (native honey) that the bees put into termite mounds (antbed). Again, his reply was vigorous: 'Ant bed. No, Latatj never like it on this Country. He put it right back this way, saltwater side. Latatj never liked it.'

The rainbird (channel-billed cuckoo, *Scythrops novaehollandiae*) assisted Latatj in pushing things away; it took back into the coastal Country those species that Latatj did not want:

> Another end [salt water side], is using different tucker from us. They got different-different tucker. That's what that rainbird pushed out. It took it back. Rainbird took it back. Latatj don't want it. Latatj don't want it, Rainbird took it back right back. That's the way. Make it good, you know.

Jimmy would always emphasise the regularities and practicalities of this organisation:

> We've got to eat the right feed that we were raised up on. That grew us up. Well, we got to go on with that same tucker ... That's why we got to go with the right feed, that we get ourselves. We might go down to Auvergne [station, saltwater zone] somewhere, I might see some tucker there, I don't know what it is. Well, I'm frightened to eat there. Very bad, you might eat some sort of tucker, poison tucker. You see, that's why that tucker that you were raised up on, that tucker, you gotta stick with that tucker. See?[14]

Nanganarri Women

The Nanganarri Women came travelling out of the west carrying tucker, future generations of people and Law. They are identified with several plant species, in particular; they deposited many foods along the way, but they also established particular associations between plants and Countries. One of their main foods is *wayita*. As discussed, *wayita* is a small edible tuber; there is an area along the Nanganarri track in Bilinara Country in which *wayita* once flourished, and the singular abundance of the tucker announced the action of the Dreaming Women.

---

14   Jimmy Manngayarri, tape 110, recorded at Yarralin, 13–14 August 1991.

Figure 4.4. *Wayita* tubers from Ngurundarni, Pigeon Hole area, 1984.
Source: Photograph by Darrell Lewis.

Ivy Kulngarri set up the photo in Figure 4.5 and captioned it: 'Two little children for Wayita Country'. The Country where the Nanganarri Women travelled and deposited the tuber *wayita* is now known as Wayita Country. The Nanganarri Women also deposited mussels in some of the billabongs, and themselves turned into nutwood trees (*partiki, Terminalia arostrata*) in the vicinity of Pigeon Hole settlement. These Dreaming Women brought 'babies' who constitute the future Bilinara generations, and they deposited them, too. There is, thus, a convergence of nurturance: both the people to take care of the Country, and the foods to take care of the people were deposited by Dreaming.

Figure 4.5. Two little children from Wayita Country, Ngurundarni, Pigeon Hole area, 1984.

Source: Photograph by Darrell Lewis.

The Nanganarri Women came out of Old Jimmy's Country to the west. There they were associated with another tuber, a yam called *pikurta* (*Ipomoea costata*). *Pikurta* does not grow in Bilinara Country; it ceases just about at the boundary between Bilinara, Ngarinman and Malngin Country. Across the black soil plains of Bilinara Country the Nanganarri Women put *wayita*. When they reached their last site in Bilinara Country, they stopped, stood up and turned into trees, before going underground to continue their travels. At this site stands a small group of bullwaddy trees (*kamanji*) (see Figure 2.1). These Dreaming trees are protected: people are not allowed to cut them down, burn them or harm them in any other way. This is the most western stand of bullwaddy. The Nanganarri Women did not come up to the surface of the Earth until they were at the edge of the region where bullwaddy is common. Their travels thus link an isolated patch of bullwaddy in the inland riverine Country with the place south-east of here in the Mudbura *kaja* Country, or arid forest, where it becomes plentiful. The Dreaming track with its songs and stories thus exists as a relationship between stands of bullwaddy, and bears knowledge of that plant at a remarkably detailed level: to know the Bilinara portion of the track is to know the location of the most westward outlier of the geographically circumscribed bullwaddy.

Black-Headed Python (Walujapi)

The Black-Headed Python cut through the Stokes Range as she came travelling out of the west. She was carrying a coolamon as she travelled, and in her coolamon she carried seeds which she put here and there. She thus established certain distribution patterns. The boundary for boab trees follows her track, and she was putting those seeds here and there, but no further to the south than where she travelled. (Boabs appear to be spreading, however, and their travels may have been sped up by white settlers, many of whom planted boabs at their homesteads.) Along the southern edge of the range, nearly every boab tree is a Dreaming tree and thus is a sacred site along the python track. It seems that these are the last of the boabs she planted along her travels. As Dreaming trees, these boabs are protected: they ought not to be cut down or damaged by human action.

Figure 4.6. Jasper Gorge was created when Walujapi, the Black-Headed Python Woman, travelled across the land, 1984.
Source: Photograph by Darrell Lewis.

The list of plants that people asserted to have been carried and deposited by the python include some which are local identifiers because of their localised distribution, and some which are local identifiers because they grow prolifically in the area. *Walmat,* for example, is a palm tree with restricted distribution, and thus far is documented at only a few locations, including Jasper Gorge (Brock 1988, 239) where the python travelled and deposited them.

This great Dreaming also carried human beings with her, and she distributed them, too. The linguistic boundary between Ngaliwurru and Karangpurru languages is marked by a site where she changed over from one language to another, and where she deposited a supply of people for the Country. She thus carried people and some of the foods that would nurture them and other living things in the region.

Travelling with totems

All of the geographies and taxonomies I have discussed are organised around overlapping and crosscutting connections. Nothing is connected to everything, and nothing is without connection. Relationships between humans and other species, technically known as totemic relations, show the same principles of crosscutting connectivities. Some totems are linked directly to land and thus precipitate groups along the geography of tracks and sites. Others are formed by different criteria and crosscut the Country-based totems. Like the Dreaming geography of Nina's painting (see Figure 3.9), totems are not confined to abstract structures, but rather connect living things across species boundaries and implicate them in relationships of care. Then, too, totems connect with seasons, growth and the contingent becoming of life in Country.

It is instructive first to look briefly at the history of the concept of totemism. Sir James Frazer's *Totemism and Exogamy* (four volumes, 1910) and Sigmund Freud's *Totem and Taboo* (1919) testify to the fact that 'totemism' was one of the cornerstones of emergent social science and related disciplines around the turn of the twentieth century. Debated regularly from decade to decade, totemism has become a palimpsest of western social theories. Definitions vary enormously, but at the core the phenomenon labelled totemism posits a non-random relationship between particular humans and particular non-humans. It is this human/non-human link that exercised the thinking of early theorists such as Frazer (1910) and Freud (1919) (discussed in Lévi-Strauss 1963, 2–3; Wolfe 1999).

This project was aimed toward distinguishing civilisation from savagery, and culture from nature. It was given special urgency by the intellectual revolution taking place in conjunction with secularisation and Darwinian theory. If humans are descended from animals, where is the boundary between them? If humans are all one family, biologically speaking, what is the difference between savagery and civilisation? These questions mattered to people who believed themselves to be fundamentally different from both animals and savages. Their project had the happy benefit of refitting under a new paradigm a set of distinctions that were both foundational and self-serving. Civilisation was marked by a separation of culture from nature, so it was said, and it followed from this that a world view that posited intimate physical relationships between people and animals must be understood as an absence of civilisation and must therefore constitute an evolutionary stage at which humans were not fully separated from nature. Totemism could thus confirm the superiority of western civilisation and the inferiority of the savage, defining and ordering their difference, while simultaneously linking them together as evolutionary moments in a global history of progress.

In 1912 Émile Durkheim wrote that 'the totem is before all a symbol, a material expression of something else. But of what?' He would go on to assert that the totem is a symbol of God and of society, brought together, in his view, in the clan (quoted in Lessa and Voigt 1979, 34). The question, 'a symbol of what?' can be read as a secular analogue to the theological longing for the absent body; it haunts social theory as a quest for the hidden referent. It proved to have an amazingly long shelf life and provided an opportunity for people to inscribe their particular theories of society and culture on the *tabula rasa* of totemism. Bronisław Malinowski, for example, in good economic fashion, found a consumption value: 'The road from the wilderness to the savage's belly and consequently his mind is very short,' he wrote in 1948, 'and for him the world is an indiscriminate background against which there stand out the useful, primarily the edible, species of plant and animal' (1948, 44). Alfred Radcliffe-Brown (1929) had developed this view in more elegant manner, suggesting that it was a common characteristic of hunting peoples to elaborate a major food item. While Radcliffe-Brown would initiate analysis into the logical properties of totems, both he and Malinowski are expressive of the theory, stated so succinctly by Claude Lévi-Strauss, that totems are 'good to eat' (Lévi-Strauss 1963, 62).

Lévi-Strauss himself found another meaning in totemism. In his view, totemism answered a universal question of the mind: 'How to make opposition, instead of being an obstacle to integration, serve rather to produce it.' Natural species, he claims, 'are chosen not because they are good to eat but because they are good to think' (1963, 80). Following the basic quest for the meaning of a totem that must be about something other than itself, Lévi-Strauss argued that totemism was about resolving problems of logic.

Lévi-Strauss's work depended on the familiar dichotomies: mind vs body, culture vs nature, difference as an obstacle to be overcome. All these points have been subjected to a range of excellent contemporary critiques (for example, Plumwood 1993). I am not proposing to go over that ground; my purpose is simply to remind readers of the culturally constructed quality of these suppositions so that it may become evident that they block access to some of the other interesting questions that could be asked.

When we bring the analytic focus into the Australian context, we see that evidence from Aboriginal Australia (especially Spencer and Gillen 1899) was drawn on by all the early theorists of totemism and 'primitive' religion. Subsequently, much of the debate about totemism was supported with evidence from Australia. Thus, virtually every major proposition concerning totemism was supported in part by reference to Australian data, and virtually every critique of theories of totemism was also supported with reference to Australian data.

Much of the anthropology of twentieth-century Australia did not seek unified global theories, but rather sought to analyse specific instances of totemic structure, action and thought. Lloyd Warner's pioneering ethnography of 1958 (1937), *A Black Civilization,* based on research he conducted in Arnhem Land in the 1920s, signals in its title the author's distance from the oppressive savagery–civilisation dichotomy. Warner stated that the totemic system of north-east Arnhem Land was 'highly elaborated and permeates all the activities of the group and all of its concepts of life in the world about it' (Warner 1958 [1937], 378). Totemism in north-east Arnhem Land, Warner contended,

> is intelligible only in terms of the social organisation, the relation of the technological system to society generally, and the ideas which surround the society's adjustment to the natural environment. (Warner 1958 [1937], 234)

In light of Warner's emphasis on both the environmental aspects of totemism and its pervasive, indeed foundational, relation to religion and social organisation, it is unfortunate that decades were to pass before these ideas were put to work systemically in other parts of the continent. Radcliffe-Brown, as stated, emphasised function, while A.P. Elkin (1933), after offering an excellent definition of totemism, sought to classify and catalogue types of totemism. In his view, totemism is formed out of relationships of mutual benefit between people and non-human species. I will return to the issue of mutual benefit in Chapter 7. W.E.H. (Bill) Stanner's 1962 phenomenological approach to totemism and religion emphasised the mystical quality of totemism (1979 [1962]). He also linked totems with clans and with Country, asserting that the group has a corporate title that covers not only the Country or site, and a mystical relation to the totemic creators, but also non-material property associated with the Country (Stanner 1965, 13). Stanner's study was closely followed by Ted Strehlow's (1970) study of Aboriginal religion in Central Australia. He documents a totemic landscape in its social, spiritual and geographical complexity. Briefly, but tantalisingly, he discussed some of the ritual which ensured the continuance of each totemic species or other existent, as had Walter Spencer and Francis Gillen (1899) before him. In this same period,

Peter Worsley's (1967) study of totemism, derived from his Groote Eylandt research, followed the tradition of Malinowski in seeking to distinguish totemism from logic and science. Nicolas Peterson (1972) followed on from Durkheim, Stanner and Strehlow in examining totemism as a link between person, group and Country. He found totemism to be a mechanism for ordering sentiment toward home place, and thus to be a key mechanism in territorial spacing (see also Strehlow 1970).

Strehlow's 1970 article 'Geography and the Totemic Landscape in Central Australia' marks a major turning point. His foundational assumption was that while totems can be and are thought to represent many things, they are also, perhaps centrally, themselves. Strehlow thus brings the living world into the analysis in a way that previous scholars, with the exception of Warner, had not done. Like many others, Strehlow agreed that the totem and the clan are connected to each other and to an area of land, and he went on to look to the organisation of ritual life oriented toward sustaining the life of the totemic species and the life of regional ecologies.[15]

The key study that initiated the analysis of totemism as ecological practice is Alan Newsome's (1980) study of the Dreaming track of the red kangaroo in Central Australia. This track traverses some of the toughest desert country in the world, and the Dreaming sites coincide with the most favoured areas for kangaroos. These sites are protected; no hunting or burning takes place in these refugia. These are places to which living things retreat during periods of stress, and from which they expand outward again during periods of plenty. Thus, kangaroos too are protected at these sites. The people responsible for that protection are the kangaroo people in whose Country the site is located.

In sum, decades of study of totemism have ensured that the clan/ Country nexus has been well analysed, but issues of human interactions and connectedness with, and responsibilities toward, the non-human world remained undervalued. Totemism posits connectedness, mutual interdependence and the non-negotiable significance of the lives of non-human species. Totemic responsibilities are organised along tracks that intersect, and thus build regional systems of relationship and responsibility (see also Rose 1997).

---

15 Following this paragraph Debbie intended to add something from the work of Morton (1990)—eds.

Back to the Victoria River

In the Victoria River District, there is a multiplicity of types of totems (discussed in greater detail in Rose 1992). Country-based totems, inherited from one's father's father, mother's father, and in many cases from mother's mother (and her brothers), link people with land, Dreaming tracks and sites, and the species of those Dreamings. The Dreaming Owlet Nightjar Jirrikit, for example, travelled and demarcated a Country and became ancestral to owlet nightjars and to the people for that Country. Jessie Wirrpa and her siblings are owlet nightjar because their father and father's father were owlet nightjars. They have responsibilities toward the sites of the owlet nightjar, and toward owlet nightjars generally, as well as owning the stories, songs, designs and sites. They are responsible for the flourishing of owlet nightjars and others of their totemic species, and this means that they are responsible for their own flourishing as well. Nor are they alone in this responsibility. Owlet nightjars (the birds) also have responsibilities toward them (the people) but it is fair to say that as a human being one knows the most about human responsibilities. Primarily, these responsibilities are to Country: to the sites, the knowledge, the practices of care and the protection of places, people and knowledge.

A Country-based totem is a singularity: you either are or are not owlet nightjar. There are other owlet nightjar clans in the world, but there is only one for this Country. This singularity is crosscut in several directions. One direction is that of other Country-based totems. Countries are exogamous, meaning that people must find their spouses in other Countries. People have non-negotiable responsibilities in their mother's father's Country as well as their father's Country, so people have totemic relationships with their mother's father's Country too. Jessie, Nina and their siblings have rights and responsibilities from their mother's father who was possum. These siblings are Countrymen with owlet nightjar, and Countrymen with possums. Countrymen take care of each other; to say that Countrymen have responsibilities toward each other's interests is to include non-humans within the realm of Law. The rule that Countries are exogamous crosscuts the singularity of Country and Country-based totems, generating kin relationships across Countries, species and people.

A system of matrilineal totems that are not connected with Country cuts across the Country-based totems. Matrilineal totems have received relatively less analysis by anthropologists, perhaps because much of the emphasis in the past few decades has been on the land tenure systems that are articulated through the system of totemic clans, but perhaps also because the real-world orientation of matrilineal totems has not been a strong focus in anthropology. Matrilineal totems in the Victoria River Country identify consubstantial relationships between persons and other species. The group of people is bounded by matrilineal descent; its members are the same flesh or 'meat' (*ngurlu*), and they share that flesh with the animal or plant with which they are associated.

As I discuss in *Dingo Makes Us Human*, these *ngurlu* groups are organised within a broader implicit organisation of matri-moieties. Within one circle of women are all the *ngurlu* that are part of dry land and the dry time of year: emu, finches, sugarleaf and *miyaka* (or *yimiyaka, Brachychiton* species). Within the other circle of women are all the *ngurlu* that are part of water, rain and the wet time of year: catfish, brolga and flying fox. Matrilineal totems thus directly concern the 'what-is' of Country and the processes that sustain it. They profoundly undercut species boundaries, and they organise groups according to ecological, not territorial, principles. They articulate patterns of connectivity such that the processes of the ephemeral world and the contingent lives of humans, animals and plants are connected in their own flesh.

Within the group of people and animals or plants who share the same flesh, what happens to one has a bearing on what happens to the other. When an emu person dies nobody eats emus until the emu people tell them they can, and similarly when a flying fox person dies nobody eats flying foxes until the right people give permission. There are more variations than there is dogma, but there is a clear recognition that the lives of same flesh beings are enmeshed in perduring relationships. Many people do not eat their own flesh. Here again, there is no dogma, with the possible exception of the emu people who neither kill nor eat emu. Many others will not kill their Dreamings but will eat them (if edible) if they are killed by others.

Genres

Luce Irigaray expands the term 'genre' to include concepts of sort, race, species and gender as well as the more familiar literary connotations of genre. Her expanded sense of the term focuses on the concept of 'kind' as in mankind, and she argues that the specificities of gender are such that women constitute their own kind or genre (Whitford 1991, 17). In my view, Irigaray's linking of 'kind' with embodied specificity gives us a way of thinking about totemism without having to invoke the whole history that emerged from Durkheim's pivotal question of what totems are really all about. They are not about absent bodies or hidden referents; they are about themselves.

The terminology for matrilineal groups in Ngaliwurru language suggests that genre is indeed an appropriate gloss. Catfish is 'fish-kind' (*yawunyung*), sugarleaf is 'plant-kind' (*mangarinyung*), emu is 'meat-kind' (*ngarinyung*) and sugarbag is honey or 'sweet-kind' (*ngalunyung*). These genres are further grouped into what one might term wet and dry genres. Flying fox and Rainbow are also grouped as light rain. Brolga and catfish together are grouped as dark rain. Emu, in contrast, is also termed *yarnanganja*— 'ground' or 'Earth' (Rose 1992, 82–83). There is no question at all of these totems representing 'something else'. Children are born from, and are the same flesh as, their mother. There is no absent body to haunt the scholarly imagination, and perhaps this is another reason why they have been undervalued by scholars.

The genre is the flesh that exists in the world. Old Tim Yilngayarri often referred to them as tribes, and he explained the origins:

> Beginning word: Dreaming been changing in that dog, changing in that crocodile ... I [Tim] came out [of] goanna Daddy, [and] catfish and brolga mother ... Born longa Namitja—brolga. We boys belong Namitja, we brolga same way, same [as] mother. That's beginning Dreaming. We different tribe, we ... And big catfish, he floats [in the water], and that brolga come along and put foot right here [on neck]. All that catfish and brolga, that's the mother and the uncle. That's all the same we tribe. Dreaming been change over ... That's we tribe.[16]

Different tribes are connected to each other; they are defined as 'mates'. They are not brothers, Old Tim said, but rather 'each one is for himself', as it follows along from mother and uncle (mother's brother).

---

16   Old Tim Yilngayarri, notebook 2, 83.

Figure 4.7. Old Tim Yilngayarri, Yarralin, c. 1981.
Source: Photograph by Darrell Lewis.

The underlying pattern has been discussed in relation to sacred geography: mutual interpenetration. Owlet nightjars, people and birds, share an essence such that humans and birds are mutually and interactively part of each other. Emus are the same, and the same relationships held between person and Country: the Country is in the person, the person is in the Country. These connectivities ramify in patterns of relationships.

Genres crosscut and overlap each other. Some of the owlet nightjar people are emu people because their mother was emu; but others are not because they had a different mother. It follows, therefore, that the members of a set of owlet nightjar people are both the same and different: the same by reason of being owlet nightjar, different by reason of their other, differing totemic relationships. As different genres crosscut each other, the people and other species who are related in one context are unrelated, or differentiated, in another context. Every boundary formed around a group of same genre folk is crosscut because people and others belong in multiple genres distinguished by context. The overlapping and crosscutting of relationships generate a mosaic of living things who are organised across species boundaries into relationships of care and interest.

Genres interrupt traditional anthropological accounts of structure and function in Aboriginal societies. From a genre perspective, the Country-based totemic clan is one among many genres rather than, as is often contended, the privileged social structure. The clan forms itself around totemic relations that are linked to Country, and the nexus of Country/Countrymen is a genre too. Footwalk epistemology is always emplaced, so the Country genre is powerfully pervasive, but it is not a singularity. Footwalk epistemology is also always on the move, and matrilineal genres may be encountered anywhere.

In addition, genres cut across western atomistic thought that would identify self or person as a singularity. The boundaries of the person are not coterminous with the body, and nor is it the case that because other people share a person's body, the person is thereby violated. On the contrary, the person achieves their maturity and integrity through relationships with people, animals, Country and Dreamings. Their being and becoming in the world exist in relation to other subjects, only some of whom are human beings. Subjectivity is not confined by the boundaries of the skin, but rather is sited both inside, on the surface of, and beyond the body. Subjects, then, are constructed both within and without; subjectivity is located within the site of the body, within the bodies of other people and other species, and within the world in trees, rock holes, on rock walls and so on. Multiple sites of subjectivity crosscut the singularities of person, species, Country and genre, and while they do not and cannot extend indefinitely, they overlap with other sites of subjectivity which are crosscut by others, which overlap with others.

Footwalk epistemology invites us to think of process rather than structure. The consubstantialities of genre form the ripples or patterns that are the created world. Action in the world is morally directed toward sustaining these relational patterns—those of difference as well as those of sameness—through the care that is attentive to embodied subjectivity.

~ ~ ~

Genres cut across contemporary social theory of self and other in a delightfully provocative way. Emmanuel Levinas is the great twentieth-century philosopher of ethical alterity. He seeks to undo western monistic philosophy 'by way of a passionate ethical protest' (quoted in Oppenheim 1997, 54). Levinas contends that the belief in the self's autonomy is a major obstacle to a life with others (Oppenheim 1997, 54), and his life's work moves away from the insular totalising self, and toward relationships. 'Self is not a substance but a relation', Levinas writes (1996, 20). He teaches an ethic of human connectivity:

> Consciousness and even subjectivity follow from, are legitimated by, the ethical summons which proceeds from the intersubjective encounter. Subjectivity arrives, so to speak, in the form of a responsibility toward an other. (Newton 1995, 12)

One of Levinas's key tropes in examining the ethical ground between self and other is the 'face'. In the concept of the face Levinas finds an unmediated ethical relation (Newton 1995, 12), a relation that calls forth subjectivity and responsibility.

The main critique that I want to bring into this dialogue originates with Luce Irigaray's 'Questions to Emmanuel Levinas' (Irigaray 1991, 109–18). Both Levinas and Irigaray agree that ethics must provide the foundation for philosophy (Oppenheim 1997, 54). Where they differ is over the specificity of the other. Levinas wants to efface difference, Irigaray maintains that erasure is violence; a system of intersubjective ethics cannot rest on erasure. To keep specificity, however, is to stay with embodiment, as in her use of the term genre to differentiate genders. Subjectivity cannot be theorised through an either–or dichotomy of substance or relationships, but rather is *both* substance and relationship.

Another of Irigaray's questions concerns 'the face of the natural universe' (Critchley 1991, 182). For Irigaray, the embodiment of human specificity is inextricable from the embodiment of the world, and elsewhere she has queried the notion of mind disengaged from nature. This approach leads her to challenge Plato's view that true selfhood is achieved through disengagement with the body and the world. She asks:

> What could induce anyone to choose as the more visible, the more true, and ultimately the more valuable something that is merely named and that is intended to replace something else that has charmed your whole life? (quoted in Plumwood 1993, 97)

Irigaray's question concerning the face of the world interrogates the gap between nature and culture but does not continue to posit a stronger theory of non-humans as intersubjective others.

Martin Buber took a similarly ambiguous stand. His 'I–Thou' is a relationship of presentness and responsibility, and although his emphasis is on the inter-human, he also considers non-humans, including nature, God and works of art (cited in Martin 1970, 244–48). I do not want to underestimate the potential for dialogue with Buber, for his is a theology of encounter and presence. It does not seek to recover, or bury, an absent body, nor does it posit that God is hidden. I do, however, want to examine the limits of Buber's theology as an ecology, because it is at that boundary that footwalk epistemology can intervene dialogically. Buber's eloquent description of encountering a tree in the 'I–Thou' mode of intersubjectivity claims for nature a reciprocal subjectivity that is not animistic: 'What I encounter is neither the soul of a tree nor a dryad, but the tree itself' (Buber 1970 [1937], 59).

Aboriginal genres intervene around substance. It is not only that self is both substance and relationships, but even more strongly that substance/self is shared across bodies and across species. And then there is place. Footwalk epistemology brings emplacement into the analysis, and thus works toward patterns of proximity and connection. Buber's hypothetical encounter between person and tree would have to become far less imaginary; it would happen in place, in time, with the histories and relationships of the parties also present to the encounter. One would want to know what kind of tree—Dreaming or ordinary, matrilineal flesh or not? And one would want to know what species, because different species have different ways of 'behaving' and thus offer their own specific beneficence in the world. All of this knowledge is part of the relationship, and thus demands the further question: who is the 'I' who encounters the tree? What is their genre, their place, their purpose?

Let us revisit the Dreaming Women and their billabong with its pandanus trees. Recall that Ivy's ethic includes non-human subjects within her world of care, communication and reciprocity. Her world is sentient, and its parts communicate. If life is always in relationship, and if communication is the evidence and much of the substance of relationships, then it follows that one of the deepest desires of all life is to be attended to, and one of the deepest practices of participation in living systems is to pay attention. If we put deep attention together with connection, we find an intersubjectivity articulated through encounter. This is a footwalk intersubjectivity. Encounter happens in the coming forth of life in the world across multiple subjectivities. It is actualised through motion, that brings living things into each other's presence.

5

'Looking at all the Country'

> When you die, you're finished up. So when you're
> alive it's good to go walking around looking at all
> the Country.
>
> Doug Campbell

Ranges

The lands that people travelled on foot are areas within which they know, with sufficient precision for survival, the waters, foods, seasons, medicines, technological items, Dreaming sites and tracks, locations of sacred and dangerous places, histories of past events, land tenure within the Aboriginal system of ownership, and land tenure and use by white settlers. If they do not hold this knowledge to begin with, they are guided and taught by those who do. Within the living areas to which people kept returning, all this knowledge was sustained at a very high degree of resolution.

Territoriality among Australian Aborigines has been subjected to excellent analysis. It is well documented, for example, that clan and tribal areas are larger in the more arid zones, and smaller in the coastal or well-watered zones. The factor is resources, and the matching of the group to a territory that will support that group. On average, group size at the 'tribal' or language level appears to have remained stable at about 300 people. A desert region with widely distributed resources would require territories that were much larger than a resource-rich territory that also supports 300 people (Birdsell 1953, 1968; Tindale 1974). There is also a good literature examining some of the social processes, such as clan fission or fusion, by which people sustain their numbers in relation to what Peterson calls their life-space (Peterson and Long 1986). These studies go to show that relationships between people and territory are, as one would expect, ecologically informed and adaptive, and indicative of long-term stability. Maps of 'tribal' boundaries demonstrate this aspect of spacing.

In this chapter I look to the footwalk maps to gain another perspective on knowledge of and relationships to eco-place. The life histories discussed above are restricted to footwalking, and thus only show a portion of people's travel and a portion of their knowledge, so it is important to add that access to vehicles has enhanced mobility greatly (Kolig 1981). Most people have driven to nearby towns such as Katherine (373 km distant) and Kununurra (381 km distant), and many have been much further afield. Hobbles, for example, went on at least one long trip associated with men's business that took him to Ringer Soak and Balgo in Western Australia. Some of the women travelled to Turkey Creek in Western Australia for women's business. Kitty Lariyari spent some years in Darwin in the leprosarium and when Cyclone Tracy hit Darwin in 1974, she and others went bush. Her knowledge of Country includes the floodplains south-east of Darwin.

Even before motor vehicles, work on the stations required mobility. In the early years women like Kitty Lariyari worked as stockmen, but that practice was discontinued (the cut-off times vary, but certainly by the 1940s most women were confined to domestic work). While women worked alongside men, they had the opportunity to learn about much the same Country as men, bearing in mind that some sites and areas are gender-specific, but after their exclusion from stock work their opportunities for learning diminished. Many of the men went on droving trips that took them as far east as Queensland, and men who worked as police 'trackers' also travelled widely in the course of their duties.

I calculated footwalk ranges for each of the participants in the map interviews (see Table 5.1). My method was to calculate square kilometre areas where it was possible to do so, as with Snowy's travels along the watershed Country. Many of the stories were presented as roads more than areas—from Centre Camp to Coolibah Station, for example. No footwalk road is ever a geometric line; it always has width as well as length. I calculated an arbitrary five-kilometre width to these roads, and then determined the area by multiplying width by distance in a straight line. My calculations underestimate areas, minimally because one does not walk in an exact straight line. I also calculated the area (roughly) in which people had lived in the bush (away from white people). It is probable that these intensive living areas have also been underestimated. Boundaries of living areas are permeable and flexible, and shifted with time, and thus are not knowable to a degree of perfect precision. I am not suggesting that the figures I produce have absolute values as footwalk ranges; they are approximations. They have all been produced by the same method, however, and thus are comparable to each other.

Table 5.1. Showing the year of birth, sex, footwalk area, intensive foraging area, extensive area, home County zone, number and type of zones visited, and the number of foods listed for each of the footwalkers discussed here.

Year of birth	M/F	Total area (km²)	Intensive (km²)	Extensive (km²)	Home zone	Number and type of zones	Foods (n)
1. 1900 Big Mick Kangkinang	M	6,960	5,040	1,920	S	3: S, R, C	73
2. 1905 Old Jimmy Manngayarri	M	9,744	3,744	6,000	R	3: S, R, D	45
3. 1907 Old Tim Yilngayarri	M	4,428	3,888	540	R	1: R	40
4. 1912 Dora Jilpngarri	F	5,400	3,600	1,800	R	2: S, R	54
5. 1913 Jambo Muntiyari	M	3,096	1,296	1,800	D	3: R, D, T	39
6. 1913 Doug Campbell	M	4,908	1,008	3,900	R	2: S, R	37
7. 1917 Jessie Wirrpa	F	2,172	1,872	300	R	1: R	53
8. 1919 Kitty Lariyari	F	6,480	2,160	4,320	R	2: R, D	32
9. 1924 Daly Pulkara	M	4,024	2,884	1,140	R	2: S, R	43
10. 1926 Hobbles Danaiyarri	M	6,564	5,184	1,380	D	2: R, D	24
11. 1927 Snowy Kulmilya	M	6,588	5,148	1,440	R	2: S, R	60
12. 1928 Charcoal Winpara	M	5,484	5,184	300	D	2: R, D, T	38
13. 1939 Nancy Kurung	F	2,364	1,584	780	R	2: S, R	45
14. 1940 Jessie Kinyayi	F	1,350	750	600	R	1: R	11
15. 1942 Riley Young	M	6,024	864	5,160	R	3: S, R, D	37

Zones: S = saltwater side; R = riverine; D = desert; C = coastal; T = Tanami (sandy desert).

Source: Author's summary.

Figure 5.1. Old Jimmy Manngayarri, a great Law man and when younger a great walker, Daguragu, 1991.

Source: Photograph by Deborah Bird Rose.

The person with the greatest range is Old Jimmy. He is acknowledged throughout the region as a person of rare walking abilities—for speed and distance he was held to be without peer, and the map confirms this assessment. He loved walking, he said, and he used to walk around and look at the Country:

> I was working man. When I'm [on] holiday, going foot. Every time, no mate, travel around all over the Country. That's why I know where the place and go through any way. My fathers, my *jawaji* [mother's fathers], all bin tell me what I know. Well I follow that Law, what he tell me, well I follow that, see.[1]

Five others fall within a roughly similar scale (Big Mick, Kitty Lariyari, Hobbles, Snowy and Riley Young). Their figures show that men's travels are likely to be more extensive than women's, even without the influence of station life. The women's areas are spread from a high that is among the highest cohort to the lowest figure in Table 5.1 (Kitty Lariyari and Jessie Kinyayi). It is significant that the three smallest areal figures are women's and the three largest areal figures are men's. It is probable that this disproportion is due to the fact that in the course of making a little boy into a young man, the lad is taken travelling by men. The ceremony requires women, and women travel for young men's business, as Dora described. Ceremony is followed by travel in the company of men. I found that men did not discuss details of this aspect of their travels in mixed company, with very few exceptions. It seems most probable that some of the long trips men made when they were young were made while they were in a state of ritual seclusion. Perhaps the most telling comparison here is that between Jessie Kinyayi and Riley Young. They are brother and sister, and their areal ranges could hardly be more different. It must be noted, however, that Jessie found that she had little patience for this work, and the figures shown probably underestimate her actual travels and her actual knowledge. Within the group of participants with large footwalk areas, age is not a factor: the oldest man and youngest man had very similar areal ranges. The only woman in this cohort is neither the oldest nor the youngest participant.

The agglomerated range does not tell the whole story, however. I calculated living and travelling figures separately and have included the separate figures. It is notable that of the participants whose intensive areas are extremely high (Big Mick Kangkinang, Hobbles Danaiyarri, Snowy Kulmilya and Charcoal Winpara), all four spent substantial amounts of time living in the bush, and two have their home Countries in desert Country, where ranges are larger because resources are more widely distributed.

---

1   Jimmy Manngayarri, tape 110, recorded at Yarralin, 13–14 August 1991.

The four oldest members of the large area group had 'maps' that combined intense travel in local areas with extensive travel. Only the youngest, Riley Young, produced a map that is far more extensive than intensive. The ratio of intensive to extensive varies across all ages, so there is not a significant pattern here. However, if we examine only those people whose home Country is riverine (thus excluding the disproportionate desert areas), one sees an uneven but general decline in intensive areas. The decline is indicative of people's shift toward sedentarisation. As noted above, with access to motor vehicles, beginning in the 1960s and rapidly increasing in the 1980s, mobility has expanded while footwalking has diminished.

In Table 5.1, I also include the home zone of each participant, by which I mean the zone within which their home Country is located. I examine not only the number of zones within which a person travelled, but also exactly which zones. Almost every participant travelled by foot in at least two zones, and many travelled in three. It is significant that people's footwalk travel only took them into zones adjacent to their home zones. Thus, Big Mick, whose home Country is the saltwater side, travelled into coastal and riverine zones. The home zone for Charcoal Winpara is located in desert, and he travelled into the Tanami (full desert) and riverine zones. Most of the participants have their home Country in the riverine zone and travelled into one or two adjacent zones. No participant travelled by foot into a zone that was not adjacent to their home zone.

In the areas to which people keep returning today (a diminished but not insignificant range), knowledge includes superbly detailed information about historical events and about environmental change. I will return to this point in Chapter 8 when I look at change and consider some of the implications of how one would know Country on a tree-by-tree basis and what the implications of that knowledge are for assessing environmental change.

Where to go?

Riley Young travelled a very long distance along a roughly western axis within a riverine/savanna *kaja* domain, and he noted the similarities that enabled him to survive. He noted the distinctiveness of the northern boundary when he defined one of the southern limits of *kakawuli,* saying that from

Bullita and Jasper Gorge north there is different tucker. He contrasted this distinction with the situation from Lingara to the south-west where, in his view, it's the same language (family) and the same tucker:

> That time we been followem that creek for tucker, bush tucker. If you go away from river, you can't findem tucker. If you go followem river, you havem plenty tucker, or plenty fish, or plenty any kind, you know. Any sort of a feed.
>
> Same tucker that one now. All the same fruit we used to live [on are found in] Gurindji, Ngarinmanpuru, [and] Kartangaruru [Country], all them same feed we been tuck out. Same tucker we used to eat, and same language we talking.
>
> All the same seeds that Country. Can't make it different. All the same seed. Only this way road [to the north], he can change different-different tucker, this way longa gorge. We used to go down to Jasper Gorge way, we used to tuck out that *kakawuli*.
>
> But this part of it right up to Inverway, Wave Hill, right up to Limbunya, Kirkimbi, all round there, that's the same feed we used to living [on]. Same fruit.
>
> And, close up to Gordon Downs, just halfway to border. All the same food we been eat. Live with same fruit, *ngapin*. No different tucker. All the same fruit.
>
> I'm thinking about when I'm looking at, you know, how many miles I been travelling, my father been carry up me. I had to walk so many miles, long way. Same fruit, same everything we been tuck out.[2]

Others also expressed the importance of being taught. Here is Daly:

> Follow my father Country. Travel all the way down, footwalk. They been tell me story first. And next time I asked to went up there and find out myself what's going on and what sort of a Country.

---

2   Riley Young, tape 86, recorded at Yarralin, 24 July 1986.

Jimmy, as usual, emphasised both intergenerational learning and Dreaming Law:

> My fathers, my *jawaji*, all bin tell me what I know. Well I follow that Law, what he tell me, well I follow that, see. Because not me, not my father, only by Dream. Everything got to be by Dream. See? Every people, they got to follow by Dream. Never make it [up] himself; by Dream, by Puwarraja. We talk Puwarraja [Dreaming]. That's why every people when you travel, you travel *kalu* [footwalk], that's the same thing [as done] by Dream. Walking all over the Country.[3]

Water and Country

Almost from my first day in Yarralin in 1980 people were explaining to me the directions of water flow. Yarralin is on the bank of the Wickham River. The water flows east into the Victoria River, and Victoria flows generally northward toward the sea, inscribing a large arc as it turns toward the west to find its outlet in the Bonaparte Gulf. Just as important as the big rivers are all the little creeks. Travelling upstream along the Wickham River you go past a number of junctions where smaller creeks flow out of the ranges and into the Wickham. Each of these is the home Country for a family group, sometimes referred to in the literature as a clan. As you go upstream, you pass through a magnificent sandstone gorge. This is the area to where the Barramundi travelled and stopped, and from where it tries to get back to its mate at Yanturi. From there, the track follows the watershed ranges between the creeks. This is Daly's Country, and Daly explained that on one side the water goes back to Gordon Creek, and on the other side the water goes back to Sanford Creek. Each creek belongs to a different mob of people, and they all come together at the junction of the Wickham and Victoria Rivers.

From my research at Yarralin, I developed an expectation that understanding the flow of water would facilitate understanding relationships between people and Country, and relationships among Countries. I have been privileged to work on many Aboriginal claims to land from the driest

---

3 Jimmy Manngayarri, tape 110, recorded at Yarralin, 13–14 August 1991. Debbie had a note here saying, 'I wonder if I should put in here about what Old Jimmy and Earth tells you. And say that I may spend the rest of my life understanding the ramifications of his statement, but what interests me here is the absolute confidence with which he organises his action in response to Country's, or Earth's, call'—eds.

corners of the driest deserts to the coastal areas and offshore islands. In most of these places, gaining a sense of water flow and, if there is no flow, of the connections between waters, greatly enhances one's understanding of groups and Country.

This insight was foreshadowed by Nicolas Peterson's 1976 article on natural and cultural regions of Australia. Peterson shows that the Australian continent can be divided into a number of large cultural regions, and that these regions conform closely to natural regions defined by water catchment. The system of family group and creek catchment is recapitulated at larger scales, not only as creeks flow into rivers, but as rivers themselves join each other, or flow into the sea in close proximity to each other.

Peterson's analysis is extremely accurate with respect to the Victoria–Ord catchment. The map he produced, based on the very small amount of research that had been carried out in the region prior to the 1980s, describes a cultural region that includes all of the Country my teachers footwalked. This analysis goes a long way toward solving what for me was an interesting problem of range.

The zones I have analysed are far more extensive in an east–west direction than in a north–south direction. I had hypothesised that an analysis of ranges would show people travelling most extensively along an east–west axis. This hypothesis was not verified. Rather, most people's ranges were roughly equal along both axes, and some people's were most extensive along a north–south axis. As stated earlier, the sample is small, and the life histories are replete with a great number of variables, so it is not possible to draw firm conclusions here, but the evidence is suggestive.

My commonsense hypothesis was not proved, and so I had to ask if there are barriers to the east and west that limited footwalk? One would readily grant that there must be physical limits to the distances a person could walk, and still return home, or walk and expect to return home and resume their place in their own society. Setting aside the possibility of endless travel, there still appear to be limits that are not primarily ecological. These limits may, therefore, be cultural. Beyond one's own catchment area the languages are unknown or unfamiliar, and the further one goes from home the fewer kin or possibilities for classificatory kin are to be found. The concurrence of cultural and natural regions seems to be sustained through a system of return in which ecological and cultural discontinuities worked together to reinforce regionalism.

The counterbalance to the idea of limits is, of course, the fact that limits are regularly exceeded. Dreaming tracks are the best example, for while most Dreamings are localised, the big travelling Dreamings like the Black-Headed Python kept going.[4] In addition to the tracks and ceremonies, contiguity has its own imperatives. The people whose Country lies on the central ridge between the east–west catchment areas of the Northern Territory include Mudbura and Jingkili people. Their Dreamings and Law are connected with those of people in both catchment areas. In recent years, with the greater mobility available to Aboriginal people through access to motor vehicles and other mechanised transport, they have facilitated contacts between people who formerly were largely, if not wholly, unknown to each other. I describe one such event in *Dingo Makes Us Human* (Rose 1992, 55–56; see also 145–49, dancing at Daguragu). Here I would add that once people organised marriages and started travelling back and forth between communities as widely separated as Yarralin and Borroloola, they commented with great interest on the environmental similarity between the two zones. Each is big river Country with much the same savanna plant communities. By foot you would have to traverse a lot of dangerous *kaja*, but by motorcar you fairly leap from one open savanna woodland to a nearly identical one in another catchment.

Storage

Amongst peoples classed as hunter-gatherers, a distinction is made between those societies in which people produce for local consumption and those in which people also produce for ceremonial feasting and exchange. Aboriginal Australians are situated within the latter category (Lourandos 1997, 18–19). The evidence for food storage is now excellent; reports from across the country, including the tropics, show that food was stored both for local consumption and for large intertribal ceremonial gatherings (49–50). Storage of grass seeds in Central Australia, for example, indicates that quantities as large as one tonne (1,000 kilograms) were stored for ceremony; ceremonies were held in periods of abundance, and the stored food helped extend the period during which people could remain together (56–57).

---

4   Debbie had a note saying that she wanted to discuss Roy Wagner's work here. This is likely to have been Roy Wagner, *The Invention of Culture*, Englewood Cliffs, NJ: Prentice-Hall, 1975—eds.

In this section I look at how food storage and food trade in the Victoria River District both undermine the old stereotype about not storing, and, more importantly, how they sustained connections amongst people across zones to produce a regional system of resource utilisation. I use the past tense here because storage is now part of living in houses with refrigerators and freezers, and the bush tucker trade is very much in decline, although it is not only a thing of the past. A number of the reasons for the decline are environmental, and I will turn to them in Chapter 8.

Formerly, the shelves or cavities within caves and rock shelters were the preferred storage sites. People kept their goods at home in their own Country, where the many modes of restricting access by strangers served also to protect one's stored goods, whether they be food, objects, bones or other material. That sense of security of home was expressed by Big Mick in his telling of the story of Walujapi, the Black-Headed Python. She gathered lily corms and had a large coolamon full. She was on top of the ranges, cooking the corms, and when she left, she left the remainder there, reckoning, 'Too heavy for me. I'll have to leave 'em here on top. Because this my Country.'[5]

A variety of foods lend themselves to storage. Some animal foods were obtained in abundance and stored, but this seems to have been rare. Few Australian animals are so large that they cannot be eaten on the spot, and of the two instances in my research only one seems to have been a pre-colonisation practice. The other, storage of beef, results from the size of the beast and the amount of food that could be eaten by a small group.

One type of animal food which often was produced in abundance was fish. The use of fish traps was one method: spinifex (probably *Triodia pungens*) was rolled and pushed along a shallow waterhole to herd fish to one end. Stone fish traps were also constructed. The use of poisons was another method. In addition, the first rains wash over hot soil and run into rivers that have shrunk back to isolated waterholes. These first rains produce hot anaerobic water that stuns the fish, and people rushed to the rivers to pick up fish by hand as they swam drunkenly in the hot and muddy water.

Snowy Kulmilya described the use of fish poison and fish storage:

> Little one fish ... we call im *walpi*. *Walpi*, yeah. We got
> a big mob, that kind, oh big mob, about a hundred.
> Get the paperbark, paperbark—you put them there

---

5   Big Mick Kangkinang, notebook 21, 86.

inside, tie em up, takem bush. No matter how long
man can stay, [the fish are] all right. [It's still] Good,
dry inside there, when you openem up. Still he's good.
They killem [crush them] gottem stone, makem stew.

Even that beef. Bullock ... cuttem, puttem la
paperbark, you can takem for months or two months,
three months, still he's good. No matter he go a bit
bad inside, he's still good. He get dry now, still good.[6]

Old Jimmy offered another account of how to store fish, and also provided
a term for dried fish powder—*niyampulu*. His words are extremely similar
to Snowy's, but the interviews took place in different years and in different
communities. The similarity seems to attest to the emotion people feel in
talking about bush foods and their former practices:

And that *yawu* [fish], *yawu,* different, you know,
when im bin cook im. Dry im out that *yawu,* all
right, tie im up longa paperbark. All right, when you
dry im now, finish, all right, put im in the paperbark.
You can leave im there, might be put im in the cave.
Might be you go next year, you going over there,
he still good![7]

Plant foods produced their own abundance seasonally and were harvested at
their peak; some were then processed for storage. Sugarleaf is an important
one. This substance forms on the leaves of two eucalypts, as Dora described
(Chapter 3). The method of processing involved breaking off the branches,
drying them on a cloth, threshing them, removing the branches and
winnowing the remainder. The sugarleaf was then wetted and made into
a flour that was shaped into cakes and wrapped in cloth or paperbark.
According to Snowy, sugarleaf could be kept for years and years. Cakes were
held in storage for ceremony times, as they were a staple for feeding the
crowds who gathered for the 'business'. They constituted one of women's
main contributions to the ceremony and have now been replaced by loaves
of bread made in camp ovens.

---

6   Snowy Kulmilya, tape 90, recorded at Yarralin, 27 July 1986.
7   Jimmy Manngayarri, tape 109, recorded at Yarralin, 13 August 1991.

According to Snowy and others, other foods for storage included lily corms, grass seeds, yams and a variety of berries and 'plums'. The method with berries and plums was to pack them into a ball, wrap the ball in paperbark, and tuck it away in a cave. The berries 'weld together' and were said to keep for years and years.

According to Old Jimmy, the yam known as *pikurta* which grows in his Country could be stored without processing:

> *Pikurta:* You can take that *pikurta,* you take im, put im there, you come back next time, you got *pikurta* there. Put it in the cave. You can leave im there, cook im, leave im there when im get dry. Or, you can carry im long way, that *pikurta.*[8]

Jimmy's statement that *pikurta* can be carried for a long way suggests that *pikurta* was one of a number of foods that were traded across the region in an organised system of exchange.

Trade

Intertribal trade is known as *winan* in most of the languages of the region and is often spoken of as buying and selling in Aboriginal Pastoral English (see Rose 1991, 7–10f). *Winan* continues, but bush tucker is rarely part of it these days. Analysis of trade patterns shows that major trade routes once connected almost the whole of the continent. There is excellent evidence for trade in a few highly prized botanical resources such as the psychotropic tobacco known as *pituri* (*Duboisia hopwoodii,* Watson 1983). This highly desirable substance grew in a restricted area of south-west Queensland. It was harvested and processed by the people who owned that Country according to knowledge of the plant and the processing requirements that were the intellectual property of the groups and were not available to other groups. Other prized items with severely localised distribution include bamboo for spear shafts, confined to a few river systems in the north-west sector of the Northern Territory, and pearl shells and baler shells from the Kimberley coast.

---

8   Jimmy Manngayarri, tape 109, recorded at Yarralin, 13 August 1991.

The distribution of resources through systems of trade responds to two subsistence imperatives: to move resources to people, and to move people to resources. In the context of moving resources to people, trade depends on two key factors: the production of surplus and the existence of scarcity. That is, one utilitarian impetus to trade is that other people have things that you want, and that you have things that they want. Regional and continental trade networks moved goods across the continent: plants, stones, shells, ochres and intellectual goods such as ceremonies. W.E.H. (Bill) Stanner (1933, 172) described the trade routes as high roads of cultural influence. The localisation of vital resources is an important factor, but the creation of scarcity through social means was also significant.

I have discussed Jirrikit's travels north into the saltwater side to get wood for spears. There is a myriad of other examples of trade in the Victoria River District, and all depend on distributing across zones and regions the technological and other items that are localised within zones or regions. Bamboo (*Bambusa arnhemica*) is traded south out of the floodplains around Darwin, which is the only region in which it grows. Hardwood spears come north out of the desert; stone tools are traded in from the east and the west where there are deposits of fine quality stone; pearl shells come from the west coast of the Kimberley region, and a fine red ochre comes from the south-east via Mudbura Country. Carrying bags, known as dilly bags, woven from palm and pandanus fibre, were traded in from both the northeast and the west. Given the necessity of many of the trade items, it is clear that trade must have been reliable and predictable.

A further aspect of trade, however, was the regional specialisation that enabled different groups to claim a unique item or range of items. People in the riverine zone made spears from a local 'bamboo' which is actually a reed (*Phragmites* sp.), and they still prized the different bamboo spears that were traded from the north. They made boomerangs from a variety of hard woods even while they traded with Mudbura people for the beautiful bullwaddy boomerangs. Regional specialisation depended on local resources as well as a variety of social conventions. Boomerangs from the desert Country were carved and decorated in ways that marked them as coming from the Tanami or beyond. Some items were (and are) sung to make them more powerful, thus enhancing their value in the trade system.

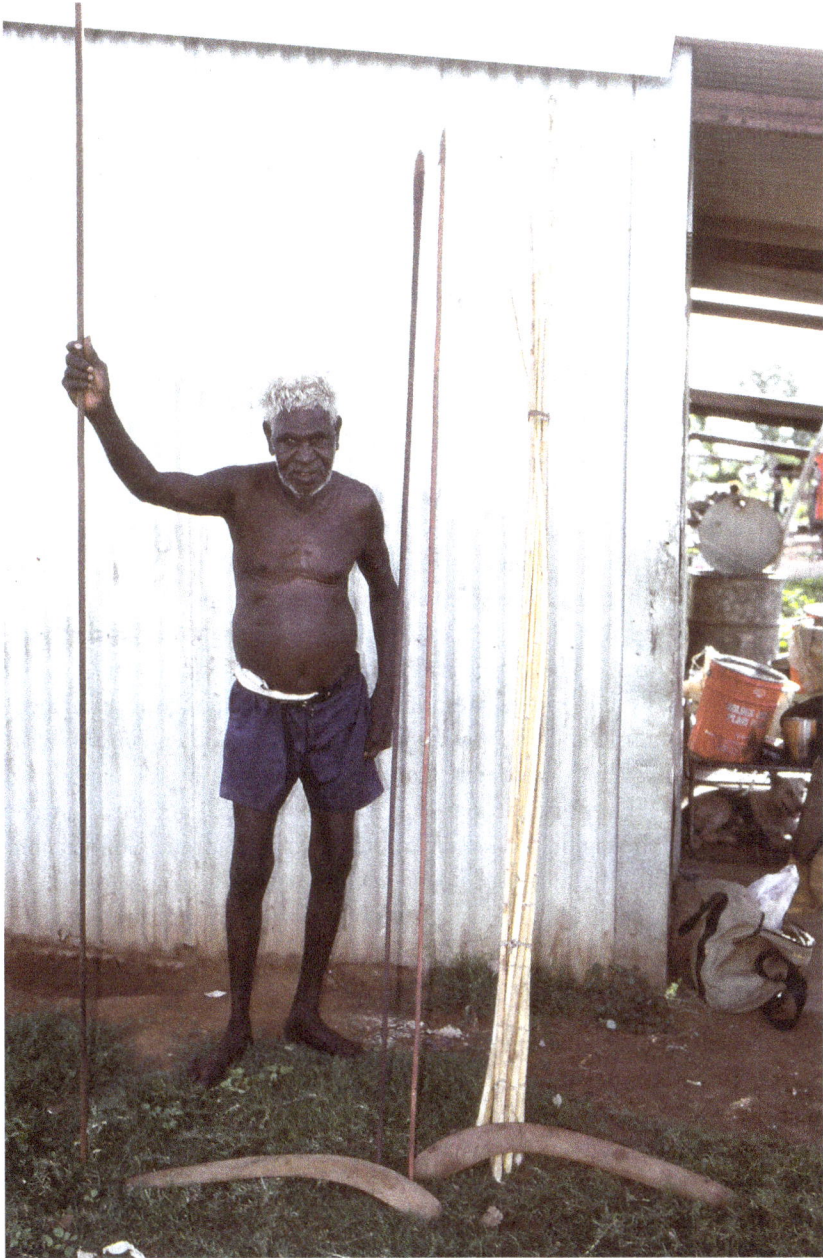

Figure 5.2. Little Mick Yinyinma with trade goods — bamboo spear shafts from the north and boomerangs from the south-east.

Source: Photograph by Darrell Lewis.

Another side of trade, far less well documented in the existing literature, is the trade in bush foods. This trade enabled surpluses to be moved around the district. Bush tucker trade is not separate from trade in other items, and like trade generally, it is productive of social relations. Riley Young explained that people brought trade goods from Wave Hill, Mount Sanford and Newcastle Waters (among other places): 'They bring and give it, we give it back tucker, spear; they bring tucker, boomerang. They call *mijelp* [themselves] relations now. In fight they go together.'[9] He is showing the exchanges that underlie and articulate relationships that sustain allies and require them to back each other up in disputes and in warfare (see Rose 1991, 101–12 for 'blackfella wars').

The foods that can be traded across distance are those that can be stored. In addition, it seems that foods achieve value when they are traded out from their own places of Dreaming origin. Sugarleaf is one such food; it can be stored in the form of cakes or loaves, and thus it can be transported for trade. Jessie Wirrpa explained: 'You can keep im months and months, sugarleaf. In a basket or anything. You keep it long time. No matter how far you take it.'[10] Hobbles was even more emphatic: 'Oh Christ, you can keep him hell of a long time.'[11] The sugarleaf that comes out of a particular Country belonging to Daly and Snowy (west of Humbert River along the watershed) is especially valued. Similarly, *miyaka,* the seeds of kurrajong (*Brachychiton*) that characterises Mudbura desert Country, was traded by the bagful and was one of several important foods that Mudbura people provided to trade partners across the region. *Miyaka* is emblematic of Mudbura Country, but also grows prolifically in the savanna *kaja* Country to the west belonging to Malngin people. According to Daly, bags of *miyaka* were sent up out of the Malngin *kaja* Country.

The seed-bearing grass of the Mudbura plains provided enormous surpluses. These areas could only ever support low population densities because of the limitations of water, but there were times and places where huge lakes formed and people could gather in large numbers. When the food in one place was far in excess of what the local group could consume it was harvested, stored for future use, and some of it was traded. Nugget Collins Ngurrartarlu said that his people used to take bags of *miyaka* and grass seeds (wild rice) to Pigeon Hole when they went footwalking there on holiday.

---

9   Riley Young, notebook 55, 97.
10 Jessie Wirrpa, tape 79, recorded at Yarralin, 15 July 1986.
11 Hobbles Danaiyarri, tape 91, recorded at Pigeon Hole, 27 July 1986.

Grass seeds flourish in different environments and ripen at different times. Lingara is the home of a Grass Seed Dreaming (and increase) site, and when the seeds became ripe there, Victoria River Downs (VRD) people who visited would bring home bags of seed.[12] Doug Campbell explained:

> Plenty longa Lingara all round there grow there ... Early days they bin always get im. All them Humbert lubra [women] bin always get im. Sometimes that VRD mob bin always go down there for holiday, you know, for Christmas, they bin always bring a pack full, all round, that *mangorlu* [grass seed]. That's goodest tucker, everything, you know. I bin eat that. My mother bin always rollem up gotta [grind] stone.[13]

Medicine was exported from the plains to the south-west of the riverine Country (from Malngin and Djaru peoples). Their Country grows a spinifex called *purrita,* and they used to gather it and smash it up and pass back to Wave Hill bags of aromatic medicine. Hobbles described the stuff:

> It doesn't grow at Yarralin, but from Daguragu Country west. It's like spinifex, only with more juice. Break it up into bits, boil it for tea, and it will bust up rubbish in the lungs. It's good for internal things. Cut it up and put it under the baby's pillow, the smell is good for him. Also, the wax from this spinifex can be boiled into tea with similar effects.[14]

Similarly, at one time the Pigeon Hole mob used to send bags of *wayita* and two different kinds of lily corms to the people at VRD Centre Camp. The lilies in the billabongs around Pigeon Hole (see Chapter 8) were there by Dreaming: some were deposited there by a Dreaming Goanna, and others were put there by the Nanganarri Women. There is an increase site there for lilies, and lilies once were extremely prolific. Their proximity to the source enhanced their value, while their abundance and storability made the trade possible.

---

12 An 'increase' site is a place dedicated to increase ceremonies, in this case for grass seed—eds.
13 Doug Campbell, tape 87, recorded at Yarralin, 25 July 1986.
14 Hobbles Danaiyarri, notebook 11, 10.

Riley Young offered a list of bush tucker trade items that includes grass seed, lily corms, bush tobacco, sugarbag, palm hearts and a berry called *tipil* (black currant), *muyin* (black plums), *martarku* (another berry, *Ziziphus quadrulocularis*) and *kumpulyu* (medicinal white currants).

The trade in bush foods was limited to the region, as far as I am able to determine. Undoubtedly distance correlates with the work of carrying bags of food or medicine, so that there are likely to be limits based on human effort. It seems also, however, that there were limits based on human knowledge. People traded in foods that they recognised and knew how to process. I asked Old Jimmy if foods came from different Countries, and he said they did not. His answer articulates a region within which bush tucker was traded:

> No, him on this Country. That's why they make im,
> sell im, just like a flour you sell im. Sell im to nother
> people. They look, 'Ah yes. Good *mangari.*' Well they
> know. Im callem name, might be sellem this way, he
> knows. 'Ah, this one so-and-so *mangari.*' He knows.
> He remember. *Mangari,* same thing. He remember.
> Like that.[15]

With the greatly increased mobility now available to Aboriginal people, the possibilities for trade are also expanded, and it is no longer the case that bush tucker trade is confined to the region. Although the trade in bush tucker has declined considerably, the interest in exotic foods has, if anything, increased. When a group of girls from Daguragu went to Ali Curung, a community in the distant desert, they brought home a bag full of *pikurta* (yams) and a bag of bush plums called *yakajiri* that were described at Daguragu as being 'like conkerberries'. The bush plum was particularly interesting to the Daguragu women because it is a tucker that is in women's business.

There is a further aspect to trade that was alluded to by Riley Young: trade is constitutive of social relations, and thus facilitates the process of moving people to resources. We have seen this in the context of ceremony and the enjoyment of abundance, but it is also significant in the context of scarcity. Richard Gould's study of the ecology of sharing emphasises the mobility of people far more than the mobility of resources, and while the emphasis may be misplaced, the argument for human mobility is valid. According to Gould (1969, 273, 1982), social relations are sustained by trade relationships, and

---

15   Jimmy Manngayarri, tape 111, recorded at Daguragu, 14 August 1991.

the need for social relations is linked to the unpredictability of Australian environments. Seasonal, annual and broader fluctuations create highly varied environments in which a resource-rich area may be ringed by areas of drought (or flood) and where this year's camp of plenty may be next year's hunger camp. Gould suggests that Aborigines stockpiled sets of social relationships based on sharing as a hedge against the uncertainties of any specific place. To quote Gould (1982, 73), 'Security in this case rests … in the widest possible set of kin relations where sharing of food and access to resources is obligatory.' Victoria River people say that *winan* (trade) is 'to make a friend together'.[16] Friends are trade partners, allies in war and refuge potentials in times of scarcity. They are bound together not only by the social interests that sustain trade, but also by the fact that trade is defined as Law. In speaking of trade, people would say: 'We run that Law right on through'. Like other forms of connectivity in motion, trading patterns crosscut, overlap and work toward mutual benefit.

Hobbles's family lived out on the grassy plains of Mudbura Country where groundwater was scarce. When the ephemeral waters dried up, and when the grass seeds were either unripe or past the point of being processed for nutrition, the families moved into other areas. Their habitation of the Country that Hobbles loved dearly was seasonal in two senses: it was contingent on water both annually and in the context of larger fluctuations. Like other people whose homelands were in the desert, they had access to several habitats, and moved in and out as the resources became available. Shortly before Hobbles was born, they left the plains Country and stayed away for a few years. In all probability, this was because of the severe droughts of the 1920s (now thought to be part of the El Niño, or southern oscillation fluctuation). The family's ability to move in and out of habitats was overtaken by European settlement. Where once they would have been with relations in known Country, conquest and the pastoralists' control of water meant that they became caught in cycles of dependency on government rations and cycles of labour on cattle stations.

These patterns of movement in and out of habitats depend on social relationships that connect people across habitats. It was almost certainly more important in the desert to be able to move in and out of eco-places, but the argument probably holds good in other zones as well, and for many of the same reasons: the unpredictability of Australian climates. The corollary

---

16   Notebook 15, 95. Debbie did not identify the person who made this statement—eds.

to leaving, of course, is the monumental abundance that occurs when the grass plains flourish with edible seeds, and every depression holds water, and the water birds arrive in the thousands.

In sum, connections across Countries and zones are sustained by travel and trade. The trade relationships make and sustain social relationships and accomplish three important ecological goals. People move themselves into areas of rich resources when reliance on their home Country resources proves inadequate, and they do this following social relations that are sustained through trade. Their presence as visitors will be reciprocated when their own periods of superabundance arrive. Second, items of limited natural distribution but wide social utility are distributed by trade. Third, local surpluses are distributed around the region by trade.

Differentiation

Zones are broader than home, but one's sense of being at home is also a sense of being in a particular kind of Country. This kind of identification is widespread across Australia and is part of a naming system in which some people are identified as sandhill people, stone Country people and the like. Home Country is associated with the specific species that distinguish it from other places. People's specific located knowledge belongs both to the people and to the Country. The desire to return home is often expressed as a longing for the specific plants or animals of home.

Along with love of home, many people also differentiate themselves from the places that are not home. This is to say that identification with place is both a positive force in its own right as well as a considered and deliberate disengagement from other places. Frequently, disengagement is expressed in relation to zones because at this level the markers of differentiation are unambiguous. Dora, characteristically, abjured any knowledge of saltwater species because of her identification as a freshwater person. We were discussing some of the saltwater species that swim the freshwater Country in the big floods:

> Sawfish? I don't know. It's for sea country, that sawfish. We don't know the name. We don't know.
>
> Debbie: Stingray?

> Dora: Oh, that one, we call it blanket. That's all we
> call it. You know, sea country we don't know names.
> Fish, you know, we don't know.

Dora then enhanced her lack of knowledge by reversing it:

> Dora: We know for sea country that *kampalngarnj.*

> Debbie: What's that?

> Dora: That fish. Belonging to the sea, where they
> always get him. They reckon he's got fat like a goanna,
> like that, this side, other side. That fish *kampalngarn,*
> they call it.

> Debbie: Might be shark?

> Dora: We don't know. That sea country fish, we don't
> know.[17]

As may be clear from Dora's account, what she did know, she did not like, and in her account of crabs she explained some of the differences to her granddaughter-in-law Ruth:

> Debbie: What about that crab? Have you eaten that
> one?

> Dora: No. Some folks eat it. They reckon it's beef
> [meat], good one. Crab. I never touched it yet, crab.
> It's too much of a cheeky one, you know. It always
> goes into a hole, like that, only it wants to frighten us.
> Cry! It made me cry! Cheeky one, crab. Olden time
> people always used to eat that crab. They reckoned it's
> good beef, crab.

> Ruth: Saltwater country got a big big one crab, eh?

> Dora: Saltwater, sea country, it's got a big one crab.
> Ordinary country, it's got a little one crab.[18]

---

17 Dora Jilpngarri, tape 82, recorded at Yarralin, 18 July 1986.
18 Dora Jilpngarri, tape 82, recorded at Yarralin, 18 July 1986.

Saltwater side people take a similarly dim view of the desert. Nancy Kurung, whose home Country was on the saltwater side, visited the savanna desert in the course of her working life as a camp cook. She described it as 'horrible country, no good, desert'. She didn't like the water, either. It was 'milky water, but it tastes ok'.[19]

Old Jimmy claimed and demonstrated a vast amount of knowledge, but he also disclaimed certain knowledge, particularly of sandhill Country. He told about Sandow, a long deceased Bilinara man, a brother to Jimmy, and like Jimmy a great walker:

> We don't know much for different-different parts of the Country, you know, and how they live. When he went down this way, my brother, old Sandow, he went down [to] Gordon Downs, Sturt Creek all around. He was living on that frog in that time. He told me about it. And he slept on a paperbark, and just covered up and slept. And he scratched the ground. *Jiljat*—that's sandy country. He made a fire and covered it up [and slept on or in the warm sand]. He lived like that, that's my brother, biggest brother, old Sandow. He lived in that Country, and he came back, right back this way, and came back right back home to Mt Sanford.[20]

Home is the place where one belongs and to which one returns, unless one's life has been subjected to great displacement. If one's life experience includes detailed knowledge of places which others do not know, a person may be somewhat stranded. Charcoal Winpara spent enough of his time in the sandhill desert Country to enable him to speak authoritatively on the different animals and plants that live there. Jambo Muntiyari also knew the sandhill desert Country, for it was his home Country. Together they spoke of many plants and animals known only to the two of them (in Yarralin). The thorny devil (a lizard, *Moloch horridus*) is a good example:

> Jambo: And nother one is *miniyiri*. It's not tucker, that one. No beef. It's poison, that one. He's got a horn like a bullock. He's got a horn like a bullock. For desert country.

---

19   Nancy Kurung, notebook 18, 41–45; Nancy Kurung, tape 29, recorded at Yarralin, 12 April 1982.

20   Jimmy Manngayarri, tape 111, recorded at Daguragu, 14 August 1991.

Charcoal: It's like a horn, he's got.

Jambo: You can't eat him.

Charcoal: He's poison.

Jambo: It's poison, that one. You only just pick him up. Anywhere in the bush they pick him up and carry him on their head. In their hair, on top, you know, they carry it into camp. And they let him go there in camp. And he's walking around, and he goes away.

Charcoal: It lives in the bush around Yuendumu country.

Jambo: All around to Hookers.

Charcoal: *Kaja* Country, all around to Hookers Creek.

Jambo: He's not for this Country. He's in my Country all around.[21]

Descriptions and lists such as Charcoal's and Jambo's demonstrate people's delight at displaying knowledge, but it is also the case that knowledge that is not shared experientially is actually of little social value. Unshared knowledge may also lead to a deep loneliness. It was no accident that Charcoal asked Jambo to come and share in the discussion of sandhill desert species. The two of them could have a conversation because they shared the life experience of that Country. For Jambo it was home Country, while for Charcoal it was Country he had lived in. Both felt at home in Country that to other people was not only strange but also from which they and their peers differentiated themselves.

---

21 Charcoal and Jambo, tape 92, recorded at Yarralin, 31 July 1986.

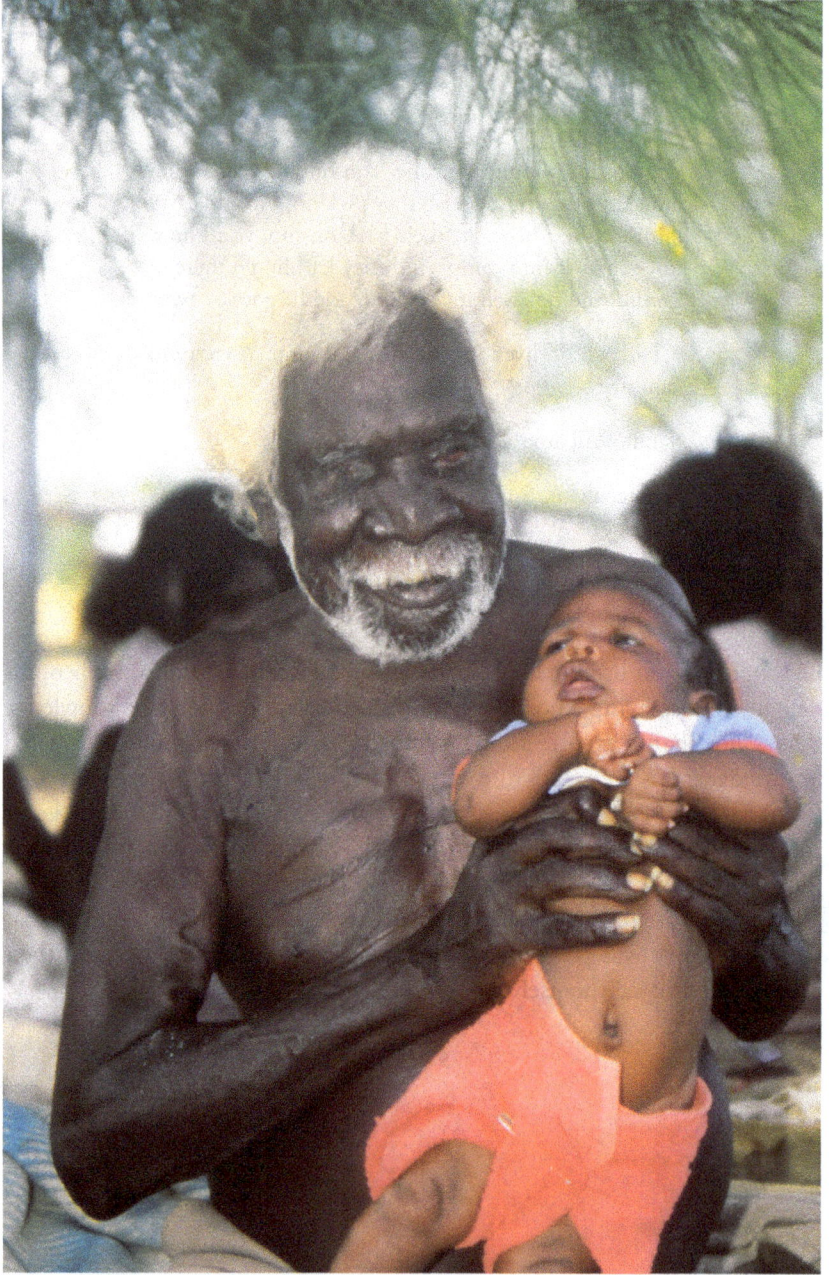

Figure 5.3. Jambo Muntiyari, Yarralin, 1982.
Source: Photograph by Darrell Lewis.

Running into change

Riley was often critical of white fellows, and he enjoyed drawing contrasts and declaring his position. He was against dams, against what NRM (natural resource management) experts would call management through the regulation of rivers. In Riley's view, the ground (which one might gloss as 'nature' or 'Earth') was a non-negotiable force in the world:

> Why that government reckon he gonna changem everything? Change him round? How you going to change him round? You can't change … that big hill there. You can't change him this ground. How you going to change him? How you going to change that creek? … Put that creek this side, he'll come back to flood this side. You can't! No way!
>
> I know government say he can change him rule. But he'll never get out of this ground. (Quoted in Rose 1992, 57)

The idea that the ground does not change is one part of the story. Another part is that everything is in a state of flux—of life and death, of nurturance, care, hunting, dying.

Change is part of life; there is nothing static in the world view of my Aboriginal teachers. They emphasise continuity and endurance, and they see these processes to be set in time and to be worked toward. Continuity, if achieved at all, is achieved against the flux and change of life and death. Sometimes continuity is not achieved. There are a few examples of Dreaming stories which seem to speak to long-term changes in the land. One such example is *miyaka*. Dreaming stories of discontinuities and plant distributions also seem to hold environmental history. *Miyaka* (a kurrajong, or *Brachychiton* species[22]) is said properly to be a desert tree. *Miyaka* Country is south of the big riverine zone and is intimately associated with groups of people for whom it is their matrilineal Dreaming and whose mothers came out of the desert (Chapter 2). And yet, a Dreaming site for *miyaka* is located in Bilinara Country, well out of the desert. Old Jimmy said that the name

---

22   Several *Brachychiton* species grow in the region, and the local Aboriginal terminologies have not adequately yet been linked to Linnaean terminologies which have recently been revised.

of this site is Miyakawurru (place for *miyaka*), and that 'Miyaka was human being before; he was walk there and turned over [into the tree]. But [there are] no *miyaka* that side'.[23]

Such stories are by no means transparent, although I believe they are suggestive of environmental histories yet to be explored. Equally, they speak to landscapes in flux, and thus to a world in which broadly defined regularities are dynamic and are subject to the work of change and transformation. Other stories concern places where trees used to be, but it is possible also to consider that some Dreaming sites, like the *miyaka* site that is not now in the immediate vicinity of *miyaka* trees, may hold a memory of a time when there were *miyaka* trees there.

In spite of the evidence in Dreaming for changes that appear to be irreversible, the model of change in people's daily lives is replacement. People die, and new generations take over. A Dreaming tree dies, and a new tree takes over. A better term than replacement is 'return'. Things happen, and life returns. An example is the tree called *yirlirli* (*Syzygium eucalyptoides*, Wightman 1994, 50). This slender tree grows in the bed of the Victoria River and in other rivers in the district. It was abundant in the Daguragu area, and people there used to call the river Yirlirlimawu. Old Jimmy remembers that he was told by old people that there was a huge flood, years and years ago, that took out all the *yirlirli*. When he was a child, he did not see *yirlirli*, and he did not learn the term until the trees started to grow again. In his view, they never came back to the full extent of their former glory. Other people do not remember the flood, and so do not compare *yirlirli* before and after. Rather, from the viewpoint of people whose memories postdate the flood, *yirlirli* is pretty much the way it has always been, bearing in mind that it fluctuates. Whether or not *yirlirli* came back to the levels it had been before the flood, it is the case that the trees have come back. Change happened, and living things reasserted their presence.

Other examples suggest that change is beneficial in establishing balance. In my travels across Humbert River Station in 1991, I saw a long strip of dead trees. On either side there were living trees, and only the one broad strip was dead. I took Riley Young up there and asked him if he had ever seen anything like that before, and what he thought of it. He said that he had seen things like that before, and that the old people said that the Rainbow Snake made that happen. The Rainbow Snake indexes underground water,

---

23   Jimmy Manngayarri, tape 110, recorded at Yarralin, 13–14 August 1991.

so this strip of dead trees was probably lying over some underground water formation. Riley's evaluation of the Rainbow Snake's action was that the trees were too crowded and too hot and so the Rainbow Snake burned them by withdrawing the water. I understand Riley's explanation to indicate self-regulation between parts of living Country: an excess of trees was curtailed by the action of the underground water.

Another type of change arises from the reflexivity between people and Country. I discussed this concept in *Dingo Makes Us Human,* focusing on the proposition that 'when old people die they kill the Country' (Rose 1992, 108). People who are Countrymen share their being with their Country, and when the Country suffers, so do people. Likewise, when people die, their Country suffers. People identify marks such as dead trees, scarred trees or scarred hills, for example, as having come into being because of the death of a person who was associated with that Country. It is the same with Dreamings: when Dreaming sites are damaged, people die; when people die their Dreamings are at risk.

Figure 5.4. Dead trees in Riley pocket, Humbert River Station, killed when the Rainbow Snake withdrew the underground water.

Source: Photograph by Darrell Lewis.

Figure 5.5. Big Mick Kangkinang and Daly Pulkara, men with tremendous knowledge of Country and Law, Yarralin, 1981.

Source: Photograph by Darrell Lewis.

When the Black-Headed Python came travelling out of Jasper Gorge, she followed a creek downstream near to where it joins the Victoria River. At one place she stopped to urinate, and a massive boab tree there always used to have water inside it that was associatively her urine. The explorer Augustus Gregory camped beneath this tree in 1856, and his party included the African explorer Thomas Baines. Baines was an artist, and he drew a sketch of this tree (see Lewis 1996, 221–25). At that time, it was intact (Figure 5.7).

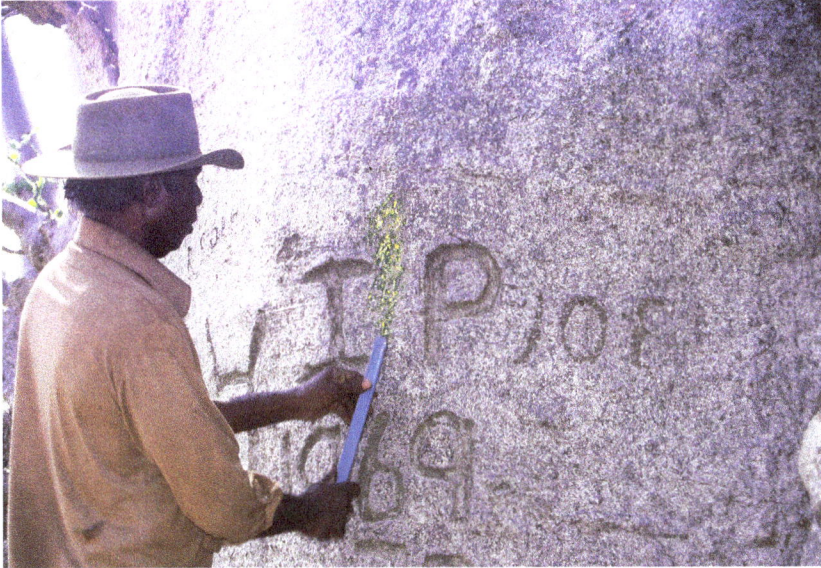

Figure 5.6. Allan Young removing a small sorcery figure from the Walujapi Dreaming boab at Wangkangki, 1986.

Source: Photograph by Darrell Lewis.

I was first introduced to the tree in 1982 when I worked with Big Mick Kangkinang and others to document sites along the track of the Black-Headed Python. We were preparing nominations for protection of these sites under the Aboriginal Sacred Sites Act of the Northern Territory. (Later, the Act was amended and retitled 'Aboriginal Areas Protection Act'.) This tree became a registered sacred site under the legislation. The tree bears the initials of white people who passed by and were moved to record their presence, and it thus holds some of the history of white people in this region, as well as the history of the Black-Headed Python, and of the people for this Country. When Jessie and Allan's father died many years ago this boab tree was said to have split in half so that all the water ran out.

In 1986 Allan Young discovered a small human figure drawing on the tree which he interpreted as a sorcery design. As a man responsible for the tree and the Country, it was his duty to remove the sorcery and, he hoped, to stop whatever illness it was causing in the world. He scraped the drawing off, and rubbed the tree with animal fat, speaking to the Black-Headed Python and telling her not to harm anyone (see Rose 1992, 66–67).

Figure 5.7. An engraving of the Black-Headed Python Dreaming boab at Wangkangki, based on a sketch that explorer Thomas Baines made in 1856.

Source: Thomas Heawood, The Boadab [i.e. Baobab] Tree [Picture] / T. Baines; T. Heawood. National Library of Australia. U208 Hand Col.; S224.; Pic Solander Box A24 #S224.

Figure 5.8. The Black-Headed Python Dreaming boab at Wangkangki in 1985, broken down but still living.

Source: Photograph by Darrell Lewis.

The creek the boab was on was cutting into the roots of the tree and threatening to undermine it. Later, in about 1990, a local white man with an interest in history and conservation decided to help preserve the tree by putting in a channel to divert the creek (Lewis 1996, 221). His actions may or may not have helped the tree; they certainly violated the Aboriginal Areas Protection Authority rules concerning work undertaken in the vicinity of a sacred site, and thus his actions were interpreted by the Aboriginal owners as hostile actions intended to harm the tree.

The tree will not live forever. It is growing out on the flats where erosion of topsoil has been extensive (Figure 5.8). People's expectation is that when the old tree dies, another tree will take its place. Given its situation in the midst of erosion, the future is uncertain.

6

Attentive Subjects

Jessie Wirrpa was a good hunter. Everybody wanted to take her on their fishing trips, or go on her fishing trips, because they knew they would be fed. Nellie Narambin and I went fishing with Jessie one day at a waterhole in the Wickham River not far from Yarralin. To the astonishment of all three of us, Jessie caught nothing. When we decided to quit, we built a little fire and boiled a billy before heading back to camp. Nellie went up to the top of the bank to wait in the truck, as she was disgusted and keen to leave. Jessie and I sat on the lower bank and drank our tea. The fire died down rapidly; we had made it from paperbark—poor firewood, but handy, and good enough for tea. The coals and ashes were still hot as we sat there drinking, and the little *nini* (zebra finches, *Taeniopygia guttata castanotis*) came down to investigate. Four or five of them gathered around the fire, moving closer and closer, until one of them pecked at a black stick. It jumped back and flew away, and the whole mob left. Then another mob came, or more probably it was the same group again. One of them pecked at a hot coal, and they rushed away again. Jessie spoke to them: 'Don't come too close, that fire still hot.' And when they came back she told them: 'Wait now, let im cool down first time.' When one of them stepped on a hot coal, leapt back and flew up into a tree, she spoke to it directly: 'I told you to wait a bit.' To me she said, 'Poor little bugger,' and she was chuckling as she spoke. They were wanting charcoal, she said, to take away to their camp.

The stillness and closeness of our bush camp became a communicative place. The way the birds would return, startle themselves and rush off, only to return again, had us laughing even before Jessie spoke to them. The moment lingers vividly in my mind. The liveliness of the world provokes an enlivened sense of one's own presence. Communication crosses intersubjective space, generating a particularly generous quality of interaction and sociality. Perhaps it does not always work this way, but that is the way it was with Jessie.

Nini are beautiful little birds, a soft fawn colour, with a black tail banded with white; their beak is orange and is circled with black. They do not speak English, it seems. If they had understood Jessie, they might have stopped trying to pick up hot coals. But to say that they do not speak English is not to say that finches do not communicate. They hang around water and can never be far from a water source. On another occasion Jessie said: 'He'll show you water. You hear im and follow im to water.' Some sources of water are very small, just a little rock hole, or a small crevice where the water flows. *Nini* are the ones who know. Snowy Kulmilya told about a time when he was walking in the bush with one old man who knew the Country. The old man said there was water there, but the ground was grassy and nothing could be seen.

> 'Yundupala [you, specify] know any water here?' That old men been tellem [asking] us. 'No. I don't know.' 'You can't see im creek here? No? Well, should be here somewhere.' When im been see im that bird *nini,* that bird [sings out] *niiii niiii* [onomatopoeia]. Well water somewhere. 'Ah, must be there' [he said]. [He] Opened up that grass, there water now.[1]

The actions of finches tell of water, and with their alert and nervy action they also tell of other presences. When emus come in for water, they stop to listen, and if the little birds are not calling out, they know they should stay away because there are people there. When the emu drinks, the little finches warn if somebody is coming. They are mates. Allan Young explained:

> That emu don't go la that water there straight away, he going to wait [till] that bird going to sing out to im. Gonna sing out la im, all right, well [that means] no man. Well im run and drink the water [then] … That's the *ninipi, nini, ninipi.* That's the bird now. He [emu's] going to wait for im … to tellem im.[2]

---

1 Snowy Kulmilya, tape 90, recorded at Yarralin, 27 July 1986.
2 Allan Young, tape 56, recorded at Yarralin, no date recorded.

On top of that, *nini* birds 'give you the time'. According to Daly Pulkara, when they start crying, that means it is time for getting emus. The hunter has to hide and wait until the *nini* tell the emu it is safe. The first time he drinks, you let him drink, Daly explained. Second time he'll shut his eye, and that's the time you kill him.[3]

When the *nini* 'give you the time', in Daly's words, they are also telling you that hot weather is here: 'Hot weather he starts crying. That's his time.' These birds are linked to the dry time of year, which is also the time when cold shifts to hot. When the shift comes, the *nini* tell you.

Hot time, cold time

In the monsoonal tropics of North Australia life responds to both sun and rain, can live without neither and is always working to avoid being overcome by one or the other. In the dry winter you can perish for lack of water; in the wet summer you can drown. Table 6.1, indicating mean rainfall over the course of the year at Victoria River Downs (VRD), shows mean variation.

Table 6.1. Mean rainfall (mm).

JAN	FEB	MAR	APR	MAY	JUN	JUL	AUG	SEP	OCT	NOV	DEC
147.8	148.6	107.0	33.3	5.9	2.1	2.9	1.1	4.5	17.5	59.6	113.6

Source: Bureau of Meteorology. Online: www.bom.gov.au, accessed 2003.[4]

Table 6.1 cannot show the great variation that occurs from year to year. Tables showing maximum and minimum monthly rainfall show the contrast (Tables 6.2 and 6.3).

Table 6.2. Highest monthly rainfall (mm).

JAN	FEB	MAR	APR	MAY	JUN	JUL	AUG	SEP	OCT	NOV	DEC
781.4	540.4	456.5	254.0	59.2	27.5	75.8	42.7	68.2	97.6	220.3	302.2

Source: Bureau of Meteorology. Online: www.bom.gov.au, accessed 2003.

Table 6.3. Lowest monthly rainfall (mm).

JAN	FEB	MAR	APR	MAY	JUN	JUL	AUG	SEP	OCT	NOV	DEC
13.4	4.9	0.0	0.0	0.0	0.0	0.0	0.0	0.0	0.0	0.0	0.0

Source: Bureau of Meteorology. Online: www.bom.gov.au, accessed 2003.

3　Daly Pulkara, notebook 37, 141–42.
4　These and other weather details are as of 2003—eds.

Because the rain is concentrated within a few months, the effects linger into the next season. In the dry season of a year following great rains your vehicle could get bogged in the soupy undersoil of the waterlogged plains, and you could die of thirst before you could get help. In the dry season following a low rainfall wet, the country becomes desiccated and remains so until the rains bring it back into growth.

Sun and rain are the two great divisions: they mark out cycles of return, and they push against each other to gain ascendancy. These are the powers that bring forth life. Rain is located at Rain Dreaming sites, and at Rainbow sites. The travelling Rain Dreamings come northwards out of the desert, and in the Victoria River Country every permanent waterhole is said to be the home of a Rainbow Snake. People usually refer to them collectively as Rainbow, and Rainbow also references most underground water. The Sun is located along the track and at the sites of its Dreaming travels, one of which is located at Yarralin.

A consideration of seasons has to start with sun and rain. Since any starting point is arbitrary, I begin in mid-year of the Anglo-Australian calendar, which is also mid-winter in the temperate zones of the southern hemisphere, and mid-dry in tropical Australia. This is the cold time of year; the sun is working, and slowly it heats up the Earth. The Earth becomes hotter and hotter, with increasing temperatures and humidity. The rain breaks the intense heat of this time of year. With the rains, the rivers start to flow again; the ephemeral waters are replenished, green grass grows and life starts to flourish. By the end of a good wet season the country is well watered and lush. The end of the rain is brought about by the big winds that bring cold weather. The sun gains ascendancy and begins to dry out the Earth and the waters, and to warm the Earth that has been cooled by rain and wind.

When I asked people about the onset of rain, I was always told that the flying foxes tell the Rainbow. In the cold time of year, the flying foxes are in the 'top country' away from the rivers. As the sun dries the country, they move toward the river, and when they get there, they hang in the trees over the river and call to the Rainbow Snake to rise up and bring rain.

This story can be fleshed out, because while there are Dreaming origins for the relationship between Rainbow and flying foxes, and while these relationships articulate some of the main social categories that bind human and animal species into groups of shared flesh (Chapter 4), there is an ecological side of the story as well. Flying foxes feed by preference on the blossoms and nectar of eucalypts. Yarralin people point especially to the inland bloodwood

(*Corymbia terminalis*) and the magnificent tree known in vernacular English as the half bark (*C. confertiflora*). Both species produce large, showy and heavily scented flowers, so they are obvious candidates for both flying fox and human attention. The large showy flowers are a vivid announcement of a more subtle process. In the Victoria River region, eucalypts flower in succession from higher ground to lower ground, which is also to say from the drier Country on the hillsides down to the riverbanks and channels. Jessie Wirrpa divided the eucalypts into those which flower in the dry time and those which flower in the rain time. *Jartpuru* (bloodwood, *C. terminalis*) and *ngurlkuku* (half bark) are among the prominent dry time flowerers, along with several other species including *yarirra* (smoke tree, *Eucalyptus pruinosa*) which grows out on the dry flats and up the lower reaches of stony hills, and *wulwaji* (coolibah) which grows around billabongs. *Timalan* (river red gum, *E. camaldulensis*) is the outstanding example of those which flower in the rain time, and the *pakali* (river melaleucas, *Melaleuca argentea* and *M. leucadendra*, paperbarks) flower then as well. Along a big river like the Wickham the banks are lined with paperbarks and river red gums. They burst into flower just before the rains come. Jessie's observation of the succession of flowering among eucalypts is, of course, consistent with the observations of ecologists who also note that the eucalypts within a given area flower at different times (Springett 1986, 148–49).

The flying foxes follow their preferred food, and it brings them to the riverside just before the rains; they forage there in the thousands. Yarralin people say that the flying foxes talk to their mate the Rainbow, telling it to move, to get up, to get to work, to bring the rain. They say that the Earth is getting too hot, that everything is too dry. We see here a sequence of communication events: the flowering of the eucalypts is triggered by each species' own internal 'time' and its response to local conditions. The flowers are food for the flying foxes, who then follow the succession of flowers. Their presence at the river to eat the flowers of *E. camaldulensis* (and melaleucas) thus constitutes an event that signals a particular moment of seasonal succession: there will not be any more eucalyptus flowers until the rains come and revitalise the Country.

Flying foxes are joined by a chorus of others, all telling the Rainbow to move and act. The Rainbow listens and rises up; it towers over the Earth, emitting lightning, thunder and rain. People, too, may add their songs to the multitude of voices demanding rain. In addition, the first rains start to create the moisture that moves back into the sky, forming more clouds and rain.

The flowering and fruiting, the new leaves and new plants are all part of the seasonal shifts, and each shift enables some further shift. Thus, when the flying foxes get to the riverside, the river gum trees and the paperbarks are flowering. The flying foxes 'tell' the Rainbow to bring rain. The rain washes the flowers off the trees and into the water. The rain also gets the rivers flowing again, after having dried back to isolated waterholes. When the rivers start to flow, the fish start to move around again, and their food, the flowers of the white gum and the paperbark are there for them. The rain cools the Earth, and the sun loses its ascendancy. Now there is the time of year, called *mayiyul* in Ngaliwurru language, when it rains and rains. *Mayiyul* builds on a root denoting the world of plants.

The end of the rainy time, like the beginning, is brought on by an accumulation of messages. A little bird (species unidentified) sings telling it to go away. Its song, '*ladawa, ladawa*', is glossed as 'go away, go away, you get away'. And the rain starts to go. At the same time, the cold wind comes up. It rises up to wrestle with the Rainbow and send it back down into the water. People may sing their songs for the wind to send the Rainbow back into the depths of the permanent water; if the Rainbow persists, the wind may break its back. Dry and cold weather return.

One way of thinking about seasons is through the contrast between wet and dry. This contrast pulls the Earth into an interactive set: water and clouds form the wet side of the story, while the dry ground and sun form the dry side of the story. Rainbow and sun, wet and dry: their interaction moves back and forth. Wet and dry underpin the matrilineal moieties, and the moieties articulate the whole social system. The matrilineally transmitted 'flesh' categories (*ngurlu*) are grouped together: on one side is sky/water, and on the other side is dry ground. Genres on each side are interconnected, and the connections are those of the world. Thus, for example, *nini* and emu are mates, and they belong to the dry ground side of the system. On the dry side, emu, also known as *yalanganja* or dry ground, is connected to sugarbag because the marks on the emu's legs look like it was bitten by bees. Similarly, the sugarleaf (also on this dry side) is connected with goannas, because the spots on the goanna are (or resemble) sugarleaf blown on the wind and stuck onto the goanna. The dry ground side is connected with the dry season, and with sun and fire. To quote Old Tim, 'When you're walking and your feet get hot, that's fire and sun.'[5] Similarly, on the water side, the different kinds of rain, the flying foxes, fish and water birds all index their time, their life processes and their people.

---

5   Old Tim Yilngayarri, notebook 4, 26.

Figure 6.1. A rock painting of the Dreaming Emu looking at the sugarbag bees biting its legs, Ngaliwurru Country, Stokes Range, 1982.

Source: Photograph by Darrell Lewis.

As a matter of Law, one should marry to the opposite kind. Jessie and her brother Allan Young are emu people, so *nini* is a mate of theirs, too. Their marriage partners are in the fish and rain matrimoiety (discussed in greater detail in Rose 1992, 74–89). Sun and rain, marriage, the fertility of the Country and the fertility of people: these relationships are all about the regeneration of the world.

The contrast between sun and rain, while clear and unambiguous in many contexts, is by no means absolute. Rather it is crosscut by other contrasts. Another way of thinking about seasons develops around the hot/cold contrast. Dora Jilpngarri explained: *parunga* is hot weather. The big wind for cold weather is called *kalajawun,* and it brings on the cold weather called *makurru.* '*Parunga, makurru, parunga, makurru,* like that all the way. No more missed im [It never misses].'[6] The contrast between hot and cold is relative; an examination of mean air temperature at 3 pm (at VRD) indicates the range (Table 6.4).

---

6   Dora Jilpngarri, notebook 52, 82.

Table 6.4. Mean 3 pm air temperature (deg C).

JAN	FEB	MAR	APR	MAY	JUN	JUL	AUG	SEP	OCT	NOV	DEC
35.9	34.4	33.7	33.5	31.2	28.5	28.3	31.0	34.8	36.9	37.1	36.3

Source: Bureau of Meteorology. Online: www.bom.gov.au, accessed 2003.

The physical sensations of coldness are not well indicated in this diagram. One might consider that the mean daily minimum temperature is around 12, 11 and 13 degrees Celsius in June, July and August, respectively (mid-winter, or mid-dry). In contrast, the mean daily minima, in December, January and February, are 24.6, 25 and 24.4 degrees Celsius respectively, or about twice the temperature of the cold time (Bureau of Meteorology n.d.). The record high temperatures in the hot time of year are well into the 40s and in mid-dry season the record low temperatures are below freezing point (Bureau of Meteorology n.d.).

Hot and cold times are indicated by constellations. The Pleiades, or Seven Sisters (Jarinkarin), tells of cold weather and an unidentified constellation, sometimes called 'seven' because of its shape, tells the hot time. When this cluster of stars gets high in the sky, that means 'it's *parunga* [hot] properly, no more go back [to cold]'.[7] Given the location of these signs, they refer to broad shifts in the Earth's orbit, and not necessarily to local phenomena, but they also provide a clear template for identifying this contrast.

The hot/cold contrast neatly intersects the wet/dry contrast, for while hot is clearly the 'summer' period and cold is the 'winter' period, the sun heats the world and the rain cools it. Hot starts after the cold of mid-year, and cold starts with the rains that cool the Earth. One effect of the contrast between hot and cold is that the apparent absolute difference between sun and rain is crosscut by another set of contrasts which shows the one building up out of the other in a more complex fashion than is suggested by a contrasting binary. Sun and rain can be thought of as opposing forces, and people's accounts of the relationships are often flavoured with a sense of aggressive encounter, for example, in saying that the winds wrestle the Rainbow out of the sky. On the other hand, and not contradictorily, rain can and does fall at any time of year (Tables 6.1, 6.2 and 6.3), and the sun shines, even if only briefly, on almost every day of the year. The two forces interpenetrate each other; their struggle for ascendancy derives as much from their entangled mutuality as from their oppositional autonomy.

---

7   Dora Jilpngarri, notebook 52, 83. Debbie intended to say more here about what the appearance of different stars and constellations signifies—eds.

Patchiness and precision

The cold/hot contrast cuts across the dry/wet contrast and thus undermines the monolithic quality of the great sun/rain dyad. It is not so much that a dualism has been cut into more finely defined parts, but rather that it has been disrupted by the flux of the world. The key point here is patchiness— parts of each big division are scattered throughout the other. Part of each is in the other, an interpenetration that refuses dualism without eroding the integrity of each. Cold can be seen to start when the rain starts to cool off the Earth (i.e. in the hottest time of the year), and so the rain of the hot time brings the conditions that will remove its necessity.

Local terminologies articulate the patchiness of rain: there is the very first rain in the hot time (*wuruwuru*), and there is the regular settled rain of the rainy time (*yipu*). There are light and dark rains, and there is the heavy rain, described by Dora: 'Rain all day, every day, make it cold, no clear, all day rain. That puts [brings on] the cold weather. That makes it come out.'[8] Then there is the cold weather rain that comes in the dry time (*kulwarang*). There is also a concept of boundedness. Hobbles explained: 'First rain brings up the grass, last rain knocks im down.'[9]

The crosscutting of the contrasts wet/dry and hot/cold is attentive to the variability and patchiness of local climates. It subverts any form of complete dominance and allows for the interpenetration of elements: rain in the dry time, patches of cold air interrupting hot weather, patches of hot drifting into the cold time.

In sum, the wet/dry and hot/cold contrasts offer two underlying frames of reference. Wet/dry can be imagined as a digital system that alternates between the Sun and the Rain(bow). When one is ascendant the other is not, and vice versa. Hot and cold are marked by constellations and thus have external indicators, but they are also processes. Cold emerges from hot by the action of the rain, and hot emerges from cold by the action of the sun. Their actions are not uniformly continuous, but rather show variable and unpredictable domains of patchiness.

---

8   Dora Jilpngarri, notebook 52, 82–83.
9   Hobbles Danaiyarri, notebook 13, 74.

Tellers

How, then, do people keep track of local seasonal changes? Unlike some parts of Australia where anthropologists have mapped seasons onto a circular calendar with many named periods (Davis 1997, for example, but see also Hoogenraad and Robertson 1997 for a critique of this endeavour), Victoria River people organise their ecological knowledge within interactive local events. The events of the world tell the story of what is happening almost day by day. Billy Bunter of Daguragu used the word 'teller' to indicate species or events that tell what is happening. In discussing tellers, he indicated an underlying aspect of this system that is implicit in the statements of many of my teachers. This is that other living things have access to their own specific forms of knowledge. They act on their knowledge, and the knowledgeable and attentive person or other creature is able to know what others know. Billy Bunter explained in reference to a bird that is probably a swift:

> And when the wet season begins he flies low, but when the wet season over he flies really high now. That's the time for him feeling that air on top for that winter, you know. When that winter coming, go back low again and start to make that camp in the side [of the cliff] there.[10]

In this way experiential human knowledge is extended through knowledgeable access to the experience of other species as expressed in their action.

Other tellers speak both to timing and intensity. In the colder inland area of the river and desert zones, animals that hibernate in the cold tell by their behaviour of the imminence of cold weather. Along the upper Victoria at Daguragu ice sometimes forms on the water in the cold time. Crocodiles prepare for the cold time by filling their gut with mud. Old Jimmy explained:

> You know what sort of cold, might be proper cold, might be little bit cold. That fellow, that you see the *warritja* (crocodile), *warritja*, he eatem stone. He eat the stone, got it in his stomach. Well, sort of sleep now, bit of ice now in the water. We call *wartimiri*, that ice.[11]

---

10   Billy Bunter, tape 115, recorded by Darrell Lewis at Daguragu, 19 August 2000.
11   Jimmy Manngayarri, tape 114, recorded by Darrell Lewis at Daguragu, 19 August 2000.

Billy Bunter elaborated:

> You can see them on the bank, they feed in the mud when you come round the scrub. But sometimes, not hot weather, that only for cold weather. They feed on the bottom. Because that bottom soil, that mud, that got some sort of feed in it … They eat one time, and that last long time, until about the summer … They got the special teller there in the body that tell them when the summer come. Sometimes they can feel that water is warming up. Sometimes during day or night, they can feel that water warming up. They got that special camp there, underneath … They go into the water, but they go up under the bank … Like a cave. That's where they camping. Under the water, but they got a camp there. And they stick themselves up into the top area, and they got the dry camp in the top there somewhere. They dive down, but they don't live in the water. They go back into the dry area. Anyway, they can stay there till the winter over. But sometimes they can come out, go to the bank, and they dry themselves in the sun there. They can lie down all day, one on top of another. One sleeping here, another on top, another one on the top.[12]

Other reptiles and fish communicate cold weather by their behaviour. Catfish also eat mud. When you cut open a catfish and see mud in its guts, you know cold weather is coming. In the cold time, when food is scarce for them, they eat dirt along the side of the river. Billy Bunter explained:

> Even the catfish, when we get it from the river, when we open im up, we can tell from the guts there. Dirt there, when they eat dirt, that's the cold weather coming up. They tell you earlier [in advance][13] … Because winter time they got no choice, they have to show up and come in to the side. Because it's hard to find their feed in the winter time. The only feed they can have is that mud.[14]

---

12   Billy Bunter, tape 115, recorded by Darrell Lewis at Daguragu, 19 August 2000.
13   Billy Bunter, tape 114, recorded by Darrell Lewis at Daguragu, 19 August 2000.
14   Billy Bunter, tape 115, recorded by Darrell Lewis at Daguragu, 19 August 2000.

Snakes and goannas live in the ground over the cold weather time. Goannas are connected with lightning:

> *Lingga,* big brown snake, when they go in the hole, they dig themselves in more deeper. Like during the summer time they go in the normal hole, or they can sleep around in any tree trunk in the hole there. But sometimes, like during wet season they sleep in the hollow log. But snake doesn't sleep in the hollow during the winter. They dig deep into that soil. They get themselves more coverem up. Like goanna, they coverem up *mijelp* [themselves] cold weather. And first rain, first lightning, that's the time. Lightning strike, that's the time that goanna start to come out. First lightning. And that second lightning, that restart the goanna in another area. That first lightning that strike the ground, that make all the goanna come out. They love wet season, and they got to come out because that time it's got to start put in a bit of rain, and making a bit of green grass, and they want to feed on some grasshoppers now. Oh, cold weather time they cover good. They can sleep there for about that many months, until that winter over.[15]

Other event-messages are ordered by connections; co-occurrence or simultaneity is of the first importance. In and around Yarralin, when the fireflies arrive, the conkerberries (*Carissa lanceolata*) are ripe; when the march flies bite, the crocodiles are laying their eggs; when the cicadas sing out, the turtles are getting fat. Green flies arrive, and the bush plums (*Vitex glabrata*) are ripe. When the *jangarla* tree flowers, the barramundi are biting. When the seed pods of the *wanyarri (bauhinia* tree, *Lysiphyllum cunninghamii*) turn very dark red the cold is finished and the 'really hot' weather is here. Other tellers include the finches who cry out in the hot weather time, and the rainbirds (channel-billed cuckoos) whose call signals rain. Another little bird tells that it is time to get lilies, and still other events refer to the time when the rivers start to flow again: 'when the brolga sings out, the catfish start to move'.

---

15 Billy Bunter, tape 115, recorded by Darrell Lewis at Dagaragu, 19 August 2000.

A number of signs identify hot, or 'really hot', weather. They are indicative of a time of year when living things are stressed. 'Hot and more hot' is the Victoria River term.[16] The rivers have sunk back into isolated waterholes, many of the surface waters have dried up completely, the remaining ephemeral waters are evaporating rapidly. In this stress period the heat has become intense, but the rains have not yet started to fall. So much information clusters in this heat period that I hypothesise that this is a time when information is urgently needed on a very detailed scale. If one were making calculations about travelling and foraging in this time of year one would want to have extremely precise information. Failure to gauge the 'time' accurately could result in death. If, for example, you planned on getting to your next drink of water at a particular billabong, and arrived to find there was no water left, you could indeed die. In this pattern of seasonal mobility, the knowledge of the geography of water and of the timing of when to leave the ephemeral waters and turn toward permanent waters would be crucial.

Patterns

I have discussed two types of information: one is large scale and linked to global and celestial, not local, events. The other is small scale and local. The movement of the stars, and the changing hot and cold seasons, are significant in the broad sweep of life. In contrast, the correspondences and co-occurrences are very finely tuned.

I tracked information through the communities of Timber Creek, Yarralin and Lingara, Pigeon Hole, Daguragu, and with Mudbura people of the Murranji area. My tracking covered three ecological zones, five languages and included two language families. My first point of interest was to ask whether the same abstract system remained stable. Was there a system of major interactions, and was it filled in with precise information events? The answer to that question is yes. The next question was: to what extent does local information hold good? The answer is that it varies. Some events, like the call of the channel-billed cuckoo, are widespread throughout the whole district and beyond, but most of the information is far more localised.

---

16 In North Australian vernacular English, the terms are the 'build up' or, more vividly, the 'suicide season'.

A number of the same indicator species occur as part of the information system. March flies, for example, are widespread. At Yarralin they tell you that the crocodiles are laying their eggs. North of Yarralin, at Timber Creek in the saltwater zone, the march flies tell exactly the same thing, and they do so in Pigeon Hole, which is south of Yarralin. South of Pigeon Hole, however, at Daguragu, the flowering of the *jangarla* tree tells that the crocodiles are laying their eggs. When it flowers that is what it 'says'. In fact, at Daguragu, knowledge is very specific: not only about the eggs, but also about when they hatch. Old Jimmy explained:

> *Ngumpin* [Aboriginal, in this region] way him tell you, pretty flower tell you. Jarinkarin (Pleiades) come out, tell you what time that cold gonna finish. Well you'll see every tree got a pretty flower, well he must be come *parunga* [hot weather] now. *Parunga* now. You see. Cold weather mob all finish now, him *parunga* now. That's sort of October. And *jangarla* tree, you see the crocodile, you see when he go on the bank lay im egg. *Jangarla* tree, im pretty flower, pretty flower, pretty flower, finish! He got seed now. After flower, that seed. That means that crocodile egg got the baby now. When him got that pretty flower, egg, and pretty flower go away and he got that seed, well he got the baby now. He'll have a little crocodile now.[17]

Back at Yarralin, when this same *jangarla* tree flowers it tells you that the barramundi are biting. Allan Young explained:

> *Jangarla* tree, him belong barramundi tucker that one. Im fall down there, barramundi eat im that one. That's only tucker now for barramundi. When the *jangarla* got pretty flower, im sit down first time cold weather, when the wind comes up, im fall down, that's im tucker. You bin see em barramundi always walking in cold weather time. You can see im.[18]

At Timber Creek, also, *jangarla* flowers tell that the barramundi are swimming near the surface of the water and will be biting.

---

17 Jimmy Manngayarri, tape 114, recorded by Darrell Lewis at Daguragu, 19 August 2000.
18 Allan Young, tape 116, recorded by Darrell Lewis at Katherine, 24 August 2000.

The indicator status of the *wanyarri* tree is widespread. As in Yarralin, at Pigeon Hole, Daguragu, and east into the Mudbura desert Country when the *wanyarri* pods become a very deep red they tell you that the really hot weather is here. They darken up at different calendrical times from year to year, and the drier the Country the sooner they darken. So, while they say the same thing, they say it in response to completely local conditions, speaking precisely to the local arena, and varying in timing from place to place and from year to year.

Going south from Daguragu onto the edge of one of the great treeless plains of the desert fringe, there are no *wanyarri* trees. There, *Grevillea dimidiata* tells you that the hot weather is here. Further afield, to the north in the floodplains south of Darwin (Rose et al. 2002, 46), the fireflies tell you that goose eggs and saltwater crocodile eggs are ready to harvest (whereas in Yarralin they tell you about conkerberries), and the march flies tell you the barramundi are biting.

In sum, the rhythms of life happening simultaneously constitute crucial information for knowing what is happening in the world. Living things communicate—the stinging bite of the march fly, the sounds of cicadas and the smell of flying foxes. They communicate by their presence or absence. Absence is crucial here: events that occur to the same rhythm require intervals of non-occurrence. There are times when things do not happen, and it is the not-happening that makes it possible for the happening to have meaning. Thus, for example, the march flies announce that the crocodiles are laying their eggs; their arrival is only remarkable because for a long time they have not been around at all. Presence and absence are thus observed, remarked upon and correlated with other events. The knowledge of regularities is learned and transmitted because many presences and absences are differences of an order that makes a difference: they are information.

~ ~ ~

I tracked the question of what, if anything, tells that crocodiles are laying their eggs. From Timber Creek through Yarralin, Lingara and Pigeon Hole, the march flies are the tellers. In Daguragu the *jangarla* trees are the tellers. This differentiation brings salt and a large portion of fresh water into a single communicative code. It places the most inland segment of the river in another code. It is not a question that can define *kaja* insofar as *kaja* is defined as an absence of water. The difference thus marked is consistent with another difference: the occasional presence of ice on water.

Daguragu is in ice Country. The term for ice is *wartimiri,* and it refers to all forms of ice, including the ice that forms on small bodies of water, and that which you might find in your billy can when you get up early. Yarralin is just north of the area in which surface ice is known. In Yarralin, the same term, *wartimiri,* refers to hail.

Old Jimmy explained that crocodiles tell you how severe the cold will be:

> You know what sort of cold will come, might be truly cold, or might be a little bit cold. That fellow, when you see the crocodile, crocodile, he's eating stones. He eats the stone, got it in his stomach. Well sort of sleeping then. And [there might be a] bit of ice now in the water. We call *wartimiri.* That's ice. Oh, you, you might be have it in the billy can, you might see ice in the billycan.[19]

Similarly with catfish. As Billy Bunter explained: 'When they go on the bank they start to eat some mud. They feed in the mud. That tells is the big winter come. Once they have that mud. That feed them for longer. The feed lasts longer.'[20] The most severely cold periods are also killers: 'That *wartimiri,* it kills fish. You know that cold, you know, cold gets into fish, it kills them.'[21]

One of the main distinctions across the region in terms of temperature is animal behaviour. I have discussed the preparations crocodiles and other water creatures make to ensure their survival during the cold months. Similarly, goannas go underground during the cold months, and only return to the surface when lightning announces hot and wet weather, and the emergence of food for goannas. The Stokes Range also marks the distinction between goanna behaviour: south of the range goannas hibernate; north of it they do not.

In sum, the pattern for ice is not the same as the pattern for hibernation. The ice boundary is consistent in a general way with the riverine concept of desert, while the hibernation boundary is consistent with the range and saltwater boundary that is so distinctive and so well known.

19 Jimmy Manngayarri, tape 114, recorded by Darrell Lewis at Daguragu, 19 August 2000.
20 Billy Bunter, tape 115, recorded by Darrell Lewis at Daguragu, 19 August 2000.
21 Jimmy Manngayarri, tape 114, recorded by Darrell Lewis at Daguragu, 19 August 2000.

As I have shown, some of these communicative signs are dispersed widely, but most are highly localised. Many of the signs of concurrence vary from place to place, so that one really only knows what is happening in the places where one has the knowledge of what concurrences exist and what they mean. The fact that this system is widespread ensures that people know that there is a system. When they go beyond the bounds of their knowledge, they still know that they are in the presence of a communicative system; furthermore, they know that they do not understand it.

Knowing that they do not know what is being communicated in the world around them makes people uncomfortable. The general unease that people often say that they feel when they are out of their own ranges is brought about in part by the loss of communication. Dora explained that when she was living in Daguragu at the time when they all walked off from VRD in 1972 (Rose 1991, 225–35), she never got much bush tucker: 'We bin fright, you know. We didn't walkabout too far because we don't know that Country, you know.' Her lack of knowledge went beyond the specifics of tellers, and included lack of knowledge of resource sites, Dreaming sites and much more. Her words must also be heard as a statement of ethics: she does not claim knowledge for other people's Country, and she does not purport to have experience of Country in the absence of knowledge.[22]

One sees quite clearly the practical importance of localised and particular information:

- each specific information event is sufficiently widespread to be useful; you do not have to hang around crocodile nesting places waiting for the right moment to start collecting eggs; you can just wait till the march flies bite (and you know when they do!);
- the information is precise at a very local level;
- the linkages between ecological information and songs, designs, Dreaming stories and sites, and matrilineal and other social categories, ensure that the information is stored and transmitted along numerous pathways;
- the information is calibrated at local levels and thus can be protected; people are knowledgeable within their own Countries, less so when they travel into other areas.

---

22 Dora Jilpngarri, tape 82, recorded at Yarralin, 18 July 1986.

Pragmatics are important because they constitute life and death knowledge. To live well in a nomadic hunter-gatherer system requires that you know where your next food is going to be, and where your next water will be. The practical aspects are inseparable from geographical, epistemological and philosophical aspects.

Here and not-here

Presence (current or past) marks the world and thus becomes part of a communicative system. Consider animal tracks. If you understand tracks, you can know what happened: what animal made the track, where it was going, how long ago this happened, and where it is likely to be now. To gain such detailed information you bring a lot of knowledge to the study. The animal did not put those tracks there as a message for you. In being itself, in living its own life in the world, it marks the world with its embodied presence. You see the marks, and if you know how to follow, you might get that creature for dinner.

Everyone is familiar with the tracks on the ground that are made by animals, including birds and insects. The Ngarinman term for track is *jamana,* which also means foot. To see a track or mark is to see the imprint of that which made it. The mark references the body of its maker. It is important to note that creatures who do not have feet also make tracks. Not only snakes, as I will discuss shortly, but also plants. *Pikurta,* for example, is a 'yam' that grows in desert Country. Hobbles explained:

> Tucker longa root inside. Longa leaf, when him grown, you know. But when all this root, he got a track … That what my mother takem me, findem tucker. Big one. Long one now. No more like a potato, but long one, different like a potato, *pikurta*. But he's a long one. You try roastem longa fire. You try havem now. He's like a potato.[23]

Other marks are sounds—the call of the birds who tell you that it is hot time. Their call is a mark they make on the world in the course of being themselves. Finches are not the St Bernards of the tropics; they do not rescue imperilled travellers by calling out to them and guiding them to

---

23   Hobbles Danaiyarri, tape 91, recorded at Pigeon Hole, 27 July 1986; the 'track' is a crack in the ground above the tuber—eds.

water. Finches do what they do; as many of my teachers said, they have their own 'Law' or their own 'culture'. Their way of doing can be understood as information by others who know how the way of being a finch is patterned with other events, places and ways of doing. Smells, too, constitute marks. You might smell the flowering eucalypts before you see them, and you will almost certainly smell the flying foxes before you see them. The smell announces the presence of the source of the smell. Then again, the march fly will mark you with its annoying little bite: you are part of the world on which the march fly imprints itself in the course of its life. Of course, you have a smell as well, so when you are hunting you want to conceal your presence.

Your sounds carry meanings too; minimally your voice and the clatter you make as you walk announce your presence. Victoria River people make noise when they walk. Unless they are hunting, they want others to know what they are doing. In grassy areas they often swat the grass with a stick as they walk along so that snakes will know they are there and keep away. Similarly, fires and smoke communicate their presence, and as I have discussed earlier, people call out to Country to announce their presence.

On the other hand, silence is important for hunting. Sign language is useful, and then there are special hunting terms. Some animals do understand human language. The echidna (*Tachyglossus aculeatus*) is an example. Snowy Kulmilya explained, starting with the colloquial term 'porcupine':

> Snowy: Them porcupine? That bloke been hunting all night. Put im in a bag, oh, big mob. Some fellow know how to get that porcupine, night time, you know. People been just killem im. Walkabout all night again. They didn't have a sleep. Roastem im there. Sit down there till he get cooked. Pick im up, take im back to camp. Feed all everybody.
>
> When they want to hunting porcupine night time, they can't talk like them people, 'I'm going hunting *junkuwuru* [porcupine]'. They can't. You can't call im name. When you go hunting night time, you can't findem porcupine [if you call his name]. Porcupine might be gone bush somewhere. Inside the cave.
>
> Debbie: So what do you have to do?

> Snowy: You just gotta talk, oh I'm going walkabout *kirinjin,* they call im. That's the what-name now, porcupine hunting. You can't talk like that: '*junkuwuru.*' [That] Frighten im. Different, eh? They talk out, 'I go *kirinjin,* this one.' We go all night, you know. Full night time, you know.

> Debbie: A bit tricky.

> Snowy: Tricky, yeah. Well porcupine he's tricky too, I think. Might be, I don't know, something [out of the ordinary]. I don't [know] how he'll find out.[24]

The sounds and smells, the marks, tracks, grooves, colours, shapes and patterns are all communicative events if you know what to pay attention to; to know that you have to know where you are, and how the information works in that area.

The absence that contributes to this information is the absence of nomadics. The whole communication system is built upon observations accumulated through time of the rhythmic returns of events and their correlations with other events in the living world. Whether it is the return of the Seven Sisters to their high place in the sky, or the return of swifts to the lower reaches of the sky, the return of march flies or the return of the red flowers and seed pods of the *wanyarri* tree, the system is understood to work through regularities of motion.

In Chapter 1, I defined nomadics as the interplay of the here and the not-here. We can now see that this same nomadics accounts for the action of the sun and rain, stars and march flies. They do their work according to an ethic of return. Victoria River people worked their own lives according to the same ethic—taking notice of what was happening and fitting their lives to it. They worked themselves ever more fittingly into the nomadics of the world through the interactive work of their own knowledgeable lives.

---

24 Snowy Kulmilya, tape 90, recorded at Yarralin, 27 July 1986.

Ecosemiotics

Let us revisit the Nanganarri Dreaming Women and their billabong for a moment and recall the idea that Ivy's ethic includes non-human subjects within her world of care, communication and reciprocity. Her world is sentient, and its parts communicate. She encounters that place situated as a part of the place; she is an owner with responsibilities to care for the place, and she is a senior person who visits the place and also shares with the place a history of encounter. If life is always in encounter, and if communication is the evidence of encounter, then it follows that one of the deepest desires of all life is to be attended to, and one of the deepest practices of participation in living systems is to pay attention. Intersubjective encounter is known through communication. If we put deep attention together with connection, we find an intersubjectivity articulated through encounter. Ivy and others taught me to see and experience a world of non-appropriative connectivity.

Some of my more conservative colleagues have wondered if all these assertions of sentience and communication are not 'just' metaphors. The argument seems to run along the lines that these Aboriginal people are too sophisticated to think that places, trees, rocks, birds and a myriad other living or inhabited things really communicate, or really are sentient. I will engage this issue with a discussion of ecosemiotics and will argue that there is nothing unsophisticated about finding communication and sentience in the world of living things.

But first, a few words about metaphor. I will consider two extremes. Many analysts contend that, since the time of the Greeks, western thought has been structured around an ontological discontinuity between the ideal and real; this ontology is embedded in language as well as articulated in philosophy and theology. Metaphor is an invitation to think of one thing in terms of another thing. It thus works across a gap of difference and invites a consideration of resemblance. If metaphor is properly understood, the reader or hearer will know that A is not B, but rather that there is some instructive reason for thinking of the similarities between the two. Metaphor is a figure of speech, and it acquires its greatest power in the figurative speech of which it is a constituent part. Even to say that it is a figure of speech, however, is to allude to that pervasive ontological gap, in this case the gap between literal and figural, between the letter and the spirit, between real and the imagined. Understood in this way, metaphor also requires a transcendent subject who is able to look at both A and B and imagine a relation or similarity between them (Ankersmit 1994; Handelman 1982; Tyler 1984).

Metaphor is powerful and pervasive in western languages and logic, but it is not universally so. Other systems of logic are based primarily on metonymy and thus work with contiguity and contextuality (see Handelman 1982). Frank Zimmerman (1996, 297) notes that ancient logic and folk taxonomies share a predilection for 'connective rather than inclusive relations'. Many non-western systems of ecological thought work with concepts of connection and communication that do not require, indeed may refuse, the idea of ontological gaps between word and thing, body and mind, world and spirit.

A wholly different idea of metaphor was suggested to me by an Aboriginal colleague and teacher, Linda Payi Ford (see Rose et al. 2002). Ford uses the term metaphor to indicate a density of meanings. She uses the metaphor of an onion to describe her understanding of the term metaphor: it refers to layers and layers of meaning. Ford speaks from a world that contains layers of information and speaks to concepts that are many faceted and multi-vocal. Her usage, while idiosyncratic, captures this density of connectedness and multiplicity of domains of meanings which are characteristic of her world view (Ford, pers. comm.).

Is it then the case that Aboriginal people do not use or recognise metaphor? I have argued against the idea of metaphor as an essential quality based on an ontological gap. There is another, less metaphysical, dimension of metaphor that involves shifting meaning from one context to another without implying a gap. Most frequently I have encountered this type of shifting meanings in the context of humour. There is a playfulness in moving ideas around out of context that underpins a great deal of social life, and gives spice, laughter and often insight into daily life. Aileen Daly is Daly Pulkara's daughter. I visited her in the year after record floods and I asked her if the floodwaters had taken out trees in the Humbert River where she lives in the community of Lingara. She said that only a few old ones had been lost. Startling me with her vivid smile, she said: 'They went for joyride.'

In speaking of communicative events, I have sought to remain as faithful as possible to the meanings of my teachers. My analysis connects readily and with great pleasure to recent work in an enlarged semiotics. Social scientists, like natural scientists, have long known that language is not the only form of communication, and that non-semantic communication is found among many animals as well as humans. From a communication perspective, Aboriginal people, like other Indigenous peoples, hold interspecies communication to be a given (for example, Scott 1996; see Guss 1985 for a broad survey). They understand ecosystems as participatory and communicative systems, and this view is in many respects similar to the views currently being developed by scholars such as Jesper Hoffmeyer (1993), who contends that the whole universe is semiotically driven. He refers to our world not as a biosphere but as a semiosphere.

The philosopher Jim Cheney (1989) discusses the concept of the relationships between language and world that rests on a totalising overlay (language over world), summarising it neatly as a view that the world is language 'all the way down' (120). He contrasts this concept of language with a concept of contextualised languages that 'percolate upward through the contexts they are bringing to voice in language shaped by this percolating process'. His neat summary of this concept is that it is 'world, all the way up' (120–21). Cheney quotes Tom Jay, who claims that 'traditional cultures' such as Native American 'bridges subject and object worlds, inner and outer … Each word bears and locates our meetings with the world' (in Cheney 1989, 121).

Aboriginal philosopher Mary Graham goes further. She offers two axioms: the land is the Law; you are not alone in the world (1999, 106). She asserts that Australian Aboriginal world views do not rest on a divide between subject and object worlds. Her analysis thus proposes not only that it is world all the way up, but also that it is Law all the way up. Law can be glossed as presence in its specificity: Law is origin, and way of being in relationship, and Law is connection. My teacher Big Mick Kangkinang put it this way: 'Tree, everything, sugarbag, tucker, goanna, fish, that no more nothing—all the fish from Dreaming. Goanna, everything, all from Dreaming.'[25] He says of everything that it is 'no more nothing'. It is presence and subjectivity, all the way up.

To know that one lives within a communicative world is not to say that one understands all the communication, or that all sentient things always understand each other. Quite the opposite is true. The finches appeared not to understand Jessie, but they are not communicatively inert. In contrast,

---

25   Big Mick Kangkinang, notebook 20, 32.

porcupines have to be tricked because they do understand human language. Different species have different languages and different cultures. A way of being in the world is also termed Law.

Sometimes people speak to other creatures, but speech is not always or even often understood. Action, however, is its own communication. Action marks the world, announcing presence (or attempting to conceal it, if necessary). The presence is the communication; the action demarcates a zone of meaning. Thus, for example, animal tracks testify to presence, and as a record of action they tell a story of what the animal was and what it did.

Within this Indigenous knowledge system there is a double decentring of the human subject. First, subjectivity in the form of sentience and agency is not solely a human prerogative but is located throughout the whole Country. Intersubjectivity is an ecological domain as well as a human one, and the ethics of encounter relate to all encounters, not only human ones.

The second decentring is that the ecological system is not run by human intentional agency, but rather calls humans into relationship and into activity. A great deal of the literature on human ecological activities— resource use and resource management (to use conventional terminology here)—assumes the priority of human knowledge and human intentional action. My work with Aboriginal people challenges this priority. Rather than humans deciding to act in the world, humans are called into action by the world. This communicative system works by calling with voice, presence, smell and other means. The result is that Country, or eco-place, far from being inert, actually brings people and other living things into being, into action, into sentience itself. It is all interactive, and it is about paying attention and being responsive.

Not only humans, but other animals as well are called to action. The lightning calls the goanna into action, as we have seen. Hobbles explained that frogs, in contrast, are silenced by lightning:

> Frog, too, when it starts to rain again he'll come out asking for water. Lightning stops him up. Frogs [are] really boss for rain, but lightning stops them every one. When lightning goes away, they going to start talking again. One fellow, he's asking the boss how long [before] that rain going to start.[26]

---

26 Hobbles Danaiyarri, notebook 11, 61.

Partial knowledge: A quick trip[27]

The pragmatics of caring for Country are based on local, fine-grained knowledge that includes the connections among living things. In order to act responsibly, humans and others must be alert to the state of the systems of which they are a part. Awareness is achieved by learning a huge body of facts concerning types and behaviour of living things, ways of interpreting behaviour, basic sets of messages, geography, Dreaming Law and places, and by continually observing and assessing what is happening. Living things give out information, their actions are messages, and other living things take notice. To be alive, and to be living rightly, is to take notice of what goes on around one. In a living system, living things are connected, and because it is a system, not a random series of unconnected events, there are patterns and predictabilities. Information about how things are connected to each other, and what is supposed to happen in conjunction with what is important, is localised land-based 'Law' or 'culture' (Aboriginal English).

Information is dispersed; specifics emerge from a background of broader categories. From this perspective the world cannot be human-centred. The march flies do not tell anybody to do anything, but those who understand them know that in this region they 'say' that the crocodiles are laying their eggs. Individuals of all species know what is going on in the world. They know because being alive and conscious they are capable of knowing, and because they have learned to understand, and, from an Aboriginal perspective, failure to pay attention is either the height of arrogance or gross stupidity.

If beings are to act wisely, they must know what is happening. Knowing is not instantaneous; it develops over time and, in important issues, depends on information which is dispersed through both time and space. An event happens, but to understand it fully one must wait to see what flows from it. The process of knowing is built up over time through an assessment of contexts and perspectives. Perspective in Indigenous systems of knowledge is also dispersed. Individuals have their own personal angle of perception, their matrilineal totemic angle, and their various Country/totemic angles which tie them into other species and to the workings of the world. To be responsible as a human person requires that one learn to recognise that other perspectives exist, and that one's own wellbeing is interwoven with that of others.

---

27   A note in the manuscript indicates that Debbie intended to rework or tighten up this entire section—eds.

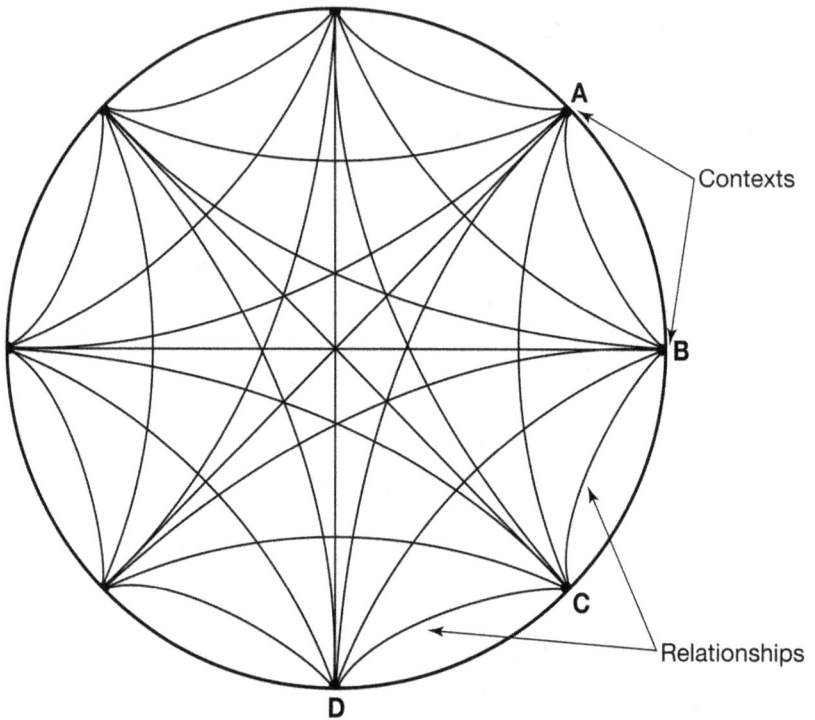

Figure 6.2. Relationships and contexts.
Source: After Figure 24, in Rose (1992, 222).

Each line in Figure 6.2 is both a boundary and a relationship. Each node (A, B, C, etc.) is both a context and an angle of vision, another centre. The view of the system changes from node to node. A wise individual (human or non-human) is capable of looking at things from several Country angles, and from the angle of various species with whom they are Countrymen or 'flesh'. No single angle defines the total, and nobody has access to every perspective. Figure 6.2 is misleading if it is read as constituting a real-life system. If this modest diagram were a real-life system, the corollary would be that there is a perspective (the reader's) from which the whole real-life system can be known. Victoria River people draw quite the opposite conclusion from their multi-centred system of dispersed knowledge: nobody knows everything, and the structure of the world is such that nobody can or should know everything.

To purport to know everything would be to undermine the integrity of the living participants in this system. The further step in this logic clearly seems to be that if the system can be known in its totality by one portion of the system (the human portion, or a segment of the human portion, let us say), then that portion can bend the system to its will, and other portions of the system cannot defend their interests because their knowledge is incomplete. Total knowledge, from this perspective, not only opens the door for a deep and enduring immorality, but also deconstructs fundamental propositions about the structure of the world:

- How can a Country and its people take care of each other if one species dominates, or if one Country dominates other Countries?
- How is mutual interdependence sustained if one group is convinced that its knowledge is greater than and encompasses the knowledge of other groups?

I have drawn out some logical consequences of the idea that a real-life world can be known fully by any segment of the world, but I must add that most of the Aboriginal people with whom I have studied do not pursue this analysis. For them it is so fundamentally obvious that the world is more complex and varied than any one angle of perception can know, and that the rights of other living things are so fundamentally part of the living world, and that a lifetime of learning leads to more questions, that their assessment of people who think their knowledge systems can, do or should encompass everything is that they are both mad and dangerous.

The morality of partial knowledge is not a completely foreign concept in western thought, as I discussed briefly in Chapter 1. It finds interesting expression in Principle 15 of the Rio Declaration (United Nations 1992): 'Where there are threats of serious or irreversible damage, lack of full scientific certainty shall not be used as a reason for postponing cost-effective measures to prevent environmental degradation.'

A more bureaucratic statement of the matter takes up this principle in the context of pollution:

> Where the state of our planet is at stake, the risks can be so high and the costs of corrective action so great, that prevention is better and cheaper than cure … where there are significant risks of damage to the environment, the government [UK] will be prepared to take precautionary action to limit the use of potentially dangerous materials, or the spread of potentially dangerous pollutants, even where scientific knowledge is not conclusive, if the balance of likely costs and benefits justifies it. (Ecologically Sustainable Development (ESD) Working Groups 1991, 41)

In my discussion with scientists about scientific practices and ethics, I have been offered more expansive views:

- If you don't have the information, don't make a decision.
- If you don't know the results of a management practice, caution and common sense indicate that you do not implement that practice.
- If you don't know what's going to happen, don't do it.
- Take action to rectify problems even without full scientific certainty.[28]

However one interprets this principle, it rests on the propositions that scientific knowledge is incomplete, that we live in a world of increasing risk in which actions have the potential to generate long-term irreversible damage, and that as a consequence of the first two propositions, caution is advisable.

The precautionary principle offers a point of contact, but where western knowledge treats lack of knowledge as an obstacle, Victoria River people treat it as an ethical situation. It goes back to absence—one's own lack of knowledge is not an empty gap to be filled, but rather an intersubjective acknowledgement that other knowledge rests with and belongs to other people and living things. Ultimately, Earth—as discussed in Chapter 9.

But locally—the person who exists in others, and in whom others exist, is vulnerable to what happens outside their own skin, but that same person finds their power in the relationships which are situated beyond the skin. To share a subjectivity is to share a self-interest. Thus, responsibilities are understood quite profoundly to be mutual and reciprocal. Genre relationships distribute subjectivity across species and Countries such that one's individual interests are held within and realised most fully in the nurturance of the interests of those with whom one shares one's being. It follows that you cannot bring yourself into being; each living thing becomes itself in the world through the work of others in whose lives its own is held. And while no individual is connected to all others, the overlap of connections sustains patterned interdependencies.

---

28   In this section, as in the section on Indigenous knowledge systems, I have not identified individuals. My intention here is to examine the properties of systems, not the individual perceptions of them.

7

Interactive Benefits

I now turn to more detailed discussion of connectivities within the living world of eco-place. I will be analysing a system in which the interconnections among living things benefit both self and others. Indeed, a system of connectivity cannot privilege self over others, as the wellbeing of self is so intertwined with that of others that nurturance of others is also nurturance of self, and nurturance of self, properly advanced, also nurtures others.

The field research that facilitated the information is presented here. My first grant for ethnobotanical research came through in 1986. I worked with botanist David Cooper. For the most part, we travelled with a few keen individuals, and we always took people who knew the Country and people who were owners of the Country (in Aboriginal Law). Jessie Wirrpa was our keenest and most knowledgeable guide and teacher. Some of the trips we made were simply opportunistic, but as we began to develop a good range of specimens, we were able to become more targeted. If we were lacking riverside plants, we proposed a day at the river. If we were lacking dry hillside plants, we proposed a trip in that direction. Eventually, we also began to target particular plants. Thus, having heard about 'native tobaccos' without encountering any, we began to ask to be taken to places where these plants would be found. In extending the research to other communities we worked out of Lingara, being guided primarily by Jessie Wirrpa and Riley Young. In Pigeon Hole we were guided and taught primarily by Ivy Kulngarri and Molly Nyurruwangali. I did some spot checks in Daguragu with Old Jimmy Manngayarri, Bulngarri and others. Across the Murranji, as stated earlier, Nugget and Long Captain guided and taught.

As we built up the collection of specimens and information, we also expanded the data by working with people back in camp who had not been on the original expedition. We went through stacks of specimens with knowledgeable older people. Usually, a group of people attended and participated. In this way we were able to document names in numerous languages, and to gain a wide variety of information.

We found that, with a few exceptions, people did not like trying to identify a specimen just by looking at it. They wanted to know the size of the plant it was taken from, where it had been growing, how old it had been and, in some cases, what other plants had been growing nearby. In addition, they wanted to know the name or names we had been given by other people. We discussed stacks of specimens with Dora (who was not able to travel), with Hobbles (who was not always available); these interviews include input from another six to ten people. In the end, we found that we had information that exceeded our expectations in every respect.

Most of our trips began with a truck; we ended up on foot for varying lengths of time depending on our targets, our teachers and everyone's interests. The longer the footwalk portion of the research the greater the amount of information, and the greater the pleasure. Footwalking holds a person open to the place and time of the world in ways that a truck can never do. The unexpected overcame us far more frequently when we were on foot than it did when we were in the truck.

Use benefits

In working through piles of specimens with teachers like Jessie, Dora and Hobbles, it became clear that people in their age range (over 60) easily identified well over 150 plants and knew of more for which I had no specimens. Some of these plants are foods, some are technological resources or medicine, some are used in ritual, some have local Dreaming significance, some are in the song cycles and some give rise to songs of more daily use. Many have technological uses, some give you the time/season and some, of course, fit into many of these categories. In discussing the plants with my teachers, both in the bush and back in camp looking at specimens, it became clear that almost every plant that has a use has a name. As we worked through so many specimens, it became clear that many of the uses that people identified are not human uses. Thus, for example, berries that are not eaten by people were identified as tucker for another species; flowers, pollen and leaves are food for others; some bushes offered shade, some trees contained nests.

In so far as this is a use-based system of nomenclature, a point I will take up later, my starting point here is that the system is not human-centred. Having interviewed human beings, however, my focus is with human use.

Multiple uses

Pampilyi (*Capparis lasiantha*) is a good plant to start with in thinking about the multiple uses and meanings of a single species. It is a scrambling vine with thorns that point backward like barbs. It grows across numerous landforms, and in the wet time it produces a sweet fruit. The fruit is edible, and is eaten by numerous species, including humans. Because of its little thorns, it is the kind of plant that clings tenaciously if you brush against it. *Pampilyi* is in a song that is part of men's business, and no more than that is known publicly. However, *pampilyi* is also a song for what is called 'love magic'. Men sing *pampilyi* and direct the song towards their desired sweetheart. The song makes the woman long for the man who sings to her, causing her to cling to him as relentlessly as the barbed vine.

Another plant, *jalartu* (*Tinospora smilacina*), is also a vine. *Jalartu* grows after the rains come and wraps itself round and round a tree trunk. This plant is associated with rain, and there is a *jalartu* song that has the power to stop rain. The fruits are forbidden for eating, but *jalartu* is medicinal. The leaves are boiled, and the warm water is used as a healing wash that is good for any sickness or pain. *Jalartu* is also part of men's song.

Like *jalartu*, the parasitic vine *jakotakota* (*Cassytha filiformis*) can be used to send away rain. Dry the stems, burn them on the fire and the smoke hunts away rain. This versatile vine can also be used as a hair dye. Burn the stems until they are black and rub them in your hair to darken it up. This plant is also medicinal and is said to be especially good for babies. Moreover, *jakotakota* grows in such thick profusion that it makes wonderful shade. It is part of the travels of the Nanganarri Women: they rested in the shade of a *jakotakota* vine as they walked across the hot black soil plains of Bilinara Country where they were putting the little tubers called *wayita*.

Ritual and Dreaming

Some plants are used in ritual. *Jakalira* (*Ventilago viminalis*) provides the wood that is made into a fire-stick and is used in ceremony: the mother of the boy being made into a young man carries the fire-stick and keeps it burning. Another plant that is used in ritual is the aromatic grass *walayi* (*Cymbopogon bombycinus*). The vernacular English name citronella grass evokes its fresh aromatic quality. This grass is broken into small pieces and used in the first ritual of childhood in which the baby is rubbed with an aromatic slurry (Rose 1992, 61–67). *Walayi* is described as 'really boss for all the baby' because its aromatic properties work to give babies life, strength and attachments to place. Other grasses can also be used in this ritual, and different Countries draw on different grasses. The evidence thus suggests that as different grasses give the infant a different smell, they thus serve to fix the child in its own Country by virtue of the smell of the air and grass of that Country.

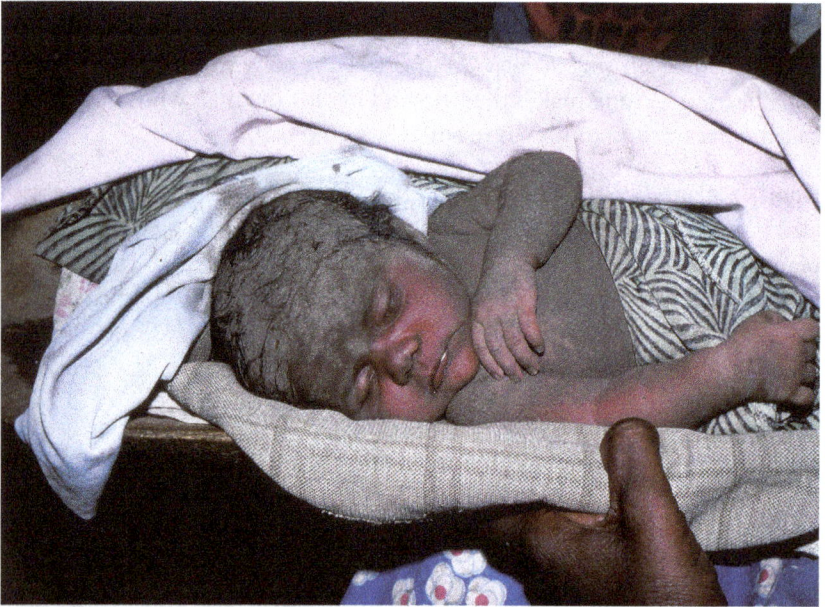

Figure 7.1. Baby (Louisa Bishop) after being rubbed with a slurry of antbed and aromatic plant material to 'fix' her to Country, Yarralin, 1982.
Source: Photograph by Darrell Lewis.

Another plant that can be used in this ritual is the *jangarla* tree (also a season marker). Women would take the bark of this tree and roast it, mix it with water and rub it on the baby, right up to the head, including the hair and all. This makes the baby 'properly black', or 'black like a crow'. Some people suggested that *jangarla* was used in trying to conceal 'half-caste' babies from the authorities who were charged with removing them from their families and placing them in institutions.[1]

As indicated, many plants have Dreaming origins, having been put in their proper place, or habitat, by travelling Dreamings. They are likely, then, to be sung when the Dreaming itself is being sung. Some plants are identifiers of regions, as I have discussed in relation to *miyaka* and savanna desert (Chapter 5). Some plants are also part of the genre system. Here again, *miyaka* is a good example. It is the matrilineal 'flesh' of several families in the region (see also 'running into change', Chapter 5). Many individual trees are also Dreaming trees. The nutwood trees at Pigeon Hole are a good example. They are Nanganarri Dreaming Women. Elsewhere, as discussed, snappy gums are uninitiated men on their travels south. Trees, generations of people and attachments to place are discussed in greater detail in Rose (1992, 106, 108, 211–13).[2]

Technologies

Technological items are numerous. Obvious ones are the straight and strong branches or saplings that are good for spears. Some are necessarily of very strong wood, so that a heavy point can be mounted on it, and it can be used to kill an animal that requires a lot of force. Strength is one factor, brittleness is another. Recall Jirrikit's efforts to get *wirlit* so that he could kill the crocodile? *Wirlit* from the north is more flexible than the inland variety. Other spears are made of less heavy material, and they are useful for hunting other animals and birds. A key property of a good spear in this monsoonal region is that the wood will hold its straight shape in the wet season. *Nyimili* (Leichhardt tree, *Nauclea orientalis*) is used for spears and spear throwers

---

1   The half-caste removal policy in the Northern Territory affected the lives of almost every family, and certainly of every community. The best analysis is *Bringing Them Home: Report of the National Inquiry into the Separation of Aboriginal and Torres Strait Islander Children from Their Families* (1997). For information on local effects see Schultz and Lewis (1995, 146–48) and Rose (1991, 169–73).
2   Debbie intended to expand this section. She had a note here observing that 'Old Jimmy Manngayarri had a strong interest in trees (along with almost everything else). Trees were people too. In creation some trees were walking around "organising" themselves'—eds.

that do not warp in the humidity. For a boomerang, you want a heavy wood that will kill on impact, and for preference you want a wood that already bends itself in the shape of boomerang. Bullwaddy (*Macropteranthes kekwickii*) is highly prized, but of course its distribution is restricted. Locally, in Victoria River Country, *Hakea arborescens* is good for boomerangs, and so is beefwood (*Grevillea striata*). For a shield, however, you want wood with flex, like the kurrajong tree (*Brachychiton diversifolius*), so that it will absorb the shock of the big fighting boomerang without shattering.

Other technological items include the wood that is used for making fire, when you have no matches (*Premna acuminata*), and the wood that catches fire so easily that it is said to be like kerosene. Another plant, curly spinifex (*Plectrachne pungens*), is used for its wax and for making windbreaks and shelters. The wax is used as a fixative in tool making. In the big river Country, sugarbag wax was preferred over both spinifex and ironwood wax, but all do the job. Yet other plants produce fibre: the cotton tree, *Cochlospermum fraseri,* for example, grows a pod with a fibrous interior not dissimilar to cotton. It was spun into string. This plant also has an edible root, although only the young plants are said to make decent eating. *Parrawi* is a fish poison and is also good for spears; in addition, its seeds are edible. Bush 'tobaccos' such as *walmalmat* (*Lobelia quadrangularis*) were important items and probably have chemical properties that produce their desired effects; they have now been largely replaced by commercial tobaccos. Other woods have other chemical properties. For chewing tobacco, you want to roll your wad of tobacco in the ashes of bark from the white gum tree (*Corymbia papuana*), and this is done with both bush and commercial tobacco.

When processing the toxic tuber *jarrwana* (*Dioscorea bulbifera*), soak it in water to which you have added ashes from the *wanyarri* tree. *Wanyarri* is also medicinal: the leaves are boiled to make a healing concoction which is drunk, and the roots can be processed to make a wash that heals sores. Native bees make their nests in *wanyarri* trees, so there may be food there. In addition, the bark and leaves can be chewed when a person is dying for water: *wanyarri* can 'save you from perishing'. One of the *wanyarri* trees not far from the Victoria River Downs (VRD) Centre Camp is a Dreaming tree: the boys who killed old Jirrikit's crocodile were swept away by the big wind, and one of them is there now, transformed into a big old *wanyarri*.

Figure 7.2. Hobbles Danaiyarri making a *nula nula* (hardwood club or fighting stick), Yarralin, 1984.

Source: Photograph by Darrell Lewis.

My emphasis is on living things, but I should mention a few others. The right kind of stone for spear points and other tools such as knives, axes and scrapers is the subject of detailed knowledge. As well, these items have been (in some cases still are) traded across long distances (Chapter 5). Today stone points are rarely used. The preferred points include wire (usually several prongs) used for fishing, hardwood, including the poisonous hardwood of the saltwater mangrove, and the big iron point called a 'shovel spear'.

There is also the body of knowledge concerning the use of animals in addition to their value as food. Kangaroo tail tendons are good for binding, and a small bone from the kangaroo makes a good tool for slicing. Animal fat is used extensively in ritual and in healing, and termite mound matter is used medicinally and in ritual (Rose 1992, 61–68).

As I have indicated, there is specific knowledge of how to process the 'cheeky' foods so that the toxins are removed, and there are detailed recipes for how to process foods, such as grass seeds or sugarleaf, to make it edible and transportable.

Medicine

Many of the medicines are fairly generalised, like the *jalartu* that you boil and bathe with and that is good for most sicknesses. However, the medicinal qualities of some plants are extremely specialised. *Kumpulyu*, or white currant (*Flueggea virosa*), is a food for people as well as a number of birds and other animals, including turkeys but not goannas. It is a 'mate' for conkerberry, and they usually ripen at about the same time. This medicine is a diuretic; the name *kumpulyu* builds on the term for urine (*kumpu*). The bark is scraped and soaked or boiled in water. The resulting drink will induce urination and reduce swelling. People said that if you've walked too much and become swollen in the legs, *kumpulyu* infusion will help. Daly reported that one man who had blood in his urine from an internal injury was treated with *kumpulyu* and recovered. Other plants are used to treat diarrhoea: *pirijpirij*, for example, produces an edible gum that is mixed with water and drunk as a tonic to strengthen digestion and to work against diarrhoea.

Lamparlampar (*Ocimum sanctum*) is a bush tea that is also medicinal. Its fragrant smell is carried on the wind at certain times of year and makes you feel good just to breathe. Some people said it was good to put under your pillow if you had a cold. If you wanted to be more direct, you could take the leaves of another medicinal and aromatic plant, *ngunungunu* (*Pterocaulon serrulatum*), and put them right in your nose. Other plants are medicinal for dogs as well as humans. *Japawin*, a fig tree, produces a milky sap that is good for treating sores on people and on dogs. At least one plant, an orchid (*Cymbidium canaliculatum*), is used to make a poison for killing dogs if they turn vicious.

Food for everybody

As I have been indicating, most plants have multiple uses. *Hakea lorea* or *pulka,* for example, is used for technology: the wood is good for boomerangs and spears. It does not produce anything edible for humans, but bees and honey-eating birds feed on the nectar of the flowers. Conkerberry (*Carissa lanceolata*) is another: the root is used to make the peg that goes on the end of a spear thrower; the wood is good for starting fires; the smoke repels mosquitoes. Almost every animal in the region eats the berries: humans, birds, dingoes, goannas, emus and turkeys. When you go walkabout for *ngamanpurru* you often see bushes where the berries have been eaten off just up to the height of a turkey, and you know who was there before you.

Muyin (bush plum) is widely beneficial. According to Dora:

> *muyin*—for kids and all. It's big tucker for that dingo,
> you know. Emu, dingo, any kind of animal tuck out
> that one. [It grows a] lotta tucker, and everyone get
> in. Bird and all, dingo and all.[3]

Jessie Wirrpa and I went for conkerberries many times. On my first expedition I was carefully picking one berry after another and putting them in my billy can with regard to avoid bruising. Having spent summers as a child in the strawberry and bean fields of Oregon I was well educated about how to pick. There I learned that you could get fired, and probably lose a lot of your earnings, if you only picked the best and biggest, or if you handled them roughly, or if you worked too fast and became careless and left some behind. Of course, you would not want to get caught eating any of the crop. I didn't think about those lessons when I was picking conkerberries with Jessie; I just picked as I knew how to pick. Before long I found that

---

3   Dora Jilpngarri, tape 74, recorded at Yarralin, 15 April 1988.

I was way out of step with the other women. They were ambling along picking the biggest, juiciest and most enticing berries and leaving all the rest behind. They ate as they picked, and they kept close together so they could chat as they moved along. Jessie brought me up and asked me why I was so slow. When I said I didn't want to waste any, she scolded me. 'Its not waste to leave im there,' she said. 'Everybody eats this tucker. This tucker for everybody. You not waste im, you leave im there for somebody nother one.'

~ ~ ~

Much of the knowledge of animal foods is extremely detailed. *Marntayark* produces fruit that is food for black cockatoos, in particular. This tree is also a major source of edible gum for humans, and the gum also has technological use as an adhesive in attaching the spear point to the spear. *Kinyuwarrangarna* (*Cyprus* sp.) is a preferred food for brolgas, and humans can eat it too. Brolgas and people eat the little bulbs that grow underground, while other birds eat the seeds when the top part flowers. This plant is strongly associated with brolgas, and thus belongs to the matrimoiety that brolgas are identified with. *Winparnin* (*Dodonaea lanceolata*) is used as a spear, as a spear point to be mounted on a different shaft and as a fighting stick (*nula nula*). This wood is said to become 'hard like steel' and can be used as a tool for flaking stone to make stone tools. Its flowers are one of the tuckers that parent birds feed to their babies. *Indigofera linifolia, karrkarta,* is a small shrub that produces little seeds. These seeds are food for small birds like the little *nini* finches and are also playthings for children; they put them in each other's hair, like nits, and pick them out.

Animals

I know less about animals than I do about plants. The mobility of most of them makes identifications more difficult, and a good number of marsupials in particular are either locally extinct or so rare as to be beyond contemporary human interaction. Unless an animal was brought into camp as food (as many were), or was relatively immobile (such as mussels), or was unmistakable (such as the sulphur-crested cockatoo), identifications were problematic.

Animal nomenclatures follow the pattern of plants. Almost every animal (mammal, bird, fish, crustacean, insect) encountered in the region bears its own name. A few animals are further divided into more detailed nomenclatures: plains kangaroo, for example, is *wawirri;* males are *jaliny;* females are *ngalijirri*. This pattern is widespread in Aboriginal Australia, although the particular species so defined vary from region to region.

The animals that occupy the daily attention of Victoria River people are foods or means to food (such as insects for bait). Like plants, many animals are part of song and ritual. Mussels, for example, were placed in the Bilinara billabongs by the Nanganarri Women. They continue to live there, doing their own dance in the water. Bilinara women have a mussel dance too, which they perform in mixed company.

Some animals are taboo to some people at certain times of their lives. Again, mussels are a good example, being taboo to young men. In addition, many animals are part of the system of genres, or 'totems', as will have become clear in previous chapters.

Mutual benefits

Many Aboriginal people in Australia articulate the idea that 'nothing is nothing' (Sutton 1988, 13). My Yarralin teachers had a slightly different take on this. Plants that to their knowledge are of absolutely no use to anyone were not named and were classed under generic terms. Most of the plants that were said to have no use were grasses and were thus classed as *yuka* (grass, generic). Dora, in particular, went on to describe these classed together plants as *walayinkarri*. Her gloss for this term was 'rubbish' or 'just the nothing'. These glosses suggest that some things (most of which turned out to be introduced grasses) really are just nothing. When I put this directly, however, asking if a specific plant was just nothing, they protested, saying, 'No, no, it's for itself'.[4]

A similar point can be made in respect of animals. As will have become clear, most animals are the descendants of Dreaming ancestors who travelled in the region. The few animals that are not part of any Dreaming are anomalies. Being out of relationship, they are out of human ken. Like plants that have no apparent use for anyone else, they exist for themselves, occupying a marginal zone of un-connection. They are just there. The only

---

4   A note in the manuscript here indicates that Debbie wanted to expand this section to include an example of a non-introduced species that is 'just for itself'—eds.

plants I can positively class as un-connected are a few grasses, some of which are introduced. The only animals I can positively class as un-connected are diverse varieties of ants.

It is interesting that cattle seem to have remained outside this system. On the one hand, there is no Dreaming for them, and they are very explicitly identified with conquest. In Hobbles's words, 'Captain Cook [came] and all his boat people, horses, cattle.' The marginal status of cattle is re-articulated through plant nomenclatures. Introduced grasses are not deemed to have a use even though they are clearly food for cattle.

The vast majority of Indigenous plants and animals are something rather than nothing, in the logic of Yarralin and Lingara people's statements. They are of use to someone other than themselves. These 'useful' species are in relationship. Their lives benefit others as well as themselves.[5]

I should like to move away from ideas of use and consider, rather, ideas of mutual benefit. Looking at species from the perspective of any individual species that is in connection, their life in the world is for both self and other. From this perspective, it is not a question of use so much as a question of benefit. The corollary is that to exist only for self is to be out of relationship, out of connectivity, because one has no benefit to offer.

As benefits move across species they start to ramify. The *Indigofera* that bears little seeds that are eaten by the *nini*/finches become implicated in the matrilineal flesh of emu people. Jessie, when she spoke to the little finches, was speaking to one of her own kind, or genre, because *nini* are mates for emus. By the same logic, *Indigofera* benefits knowledge of seasons because it benefits *nini;* it is *nini* who 'gives you the time'. And by the same token, *nini* benefits hunters like Daly who hunt the emus who drink the water that the finches mark for them by their presence. Emu people do not hunt emus, so the benefit here is not distributed across the whole of the local human species, but only across part of it (see Rose 1992, 83–84).

As I have shown, many species benefit not just one other species, but many others. *Indigofera* offers a small range of direct benefits but consider conkerberries; food for so many animals and birds, technological benefits and the glorious benefit (for humans) of repelling mosquitoes. The benefits are spread across species and the shrub itself is part of an important

---

5   Debbie intended to expand this section, possibly discussing some taxonomical studies—eds.

Dreaming story. The Emu Dreaming travelled from west to east, going through western Ngarinman Country (Bumundu Country) and Bilinara Country. She piled up conkerberries in western Ngarinman Country and the site is there today: a little hill shaped like a large pile of berries. Conkerberries nourish emus, and emus bring us back into Law, creation, matrilineal genres and land ownership.

Benefits ramify, but not forever. The pattern is by now familiar. Everything that offers benefit is in connection. Nothing is connected to everything, but as benefits ramify, crosscut, overlap and link up, they form a world of connections in which to enter that world at all is to enter dense and patterned relationships. If followed, the patterning would eventually take you through the known and named world.

Benefit in this system, then, is both direct and indirect. The reciprocity of benefit is also both direct and indirect. Birds that eat fruit and spread the seeds are an obvious example of direct reciprocity. There are far more examples of indirect reciprocity. The benefits spread widely, and the logic of the system is that as life is promoted, so the lives of living things are promoted. To put it more broadly: the promotion of relations of connectivity is a benefit in its own right, as it keeps connections flourishing, and thus remains seriously alive.

Humans engage in direct reciprocities in this system too. They are, as I have shown, the direct beneficiaries of many plants and animals. Their own reciprocal benefit is offered primarily in forms of restraint, curation and ceremony. One of the main forms of both restraint and curation is fire ecology.[6]

A fire excursion

Hobbles spoke of Aboriginal people walking around organising the Country. Fire was the major technology of organisation. The centrality of fire in Aboriginal life throughout Australia cannot be overestimated. Every European explorer from Tasmania to North Australia saw smoke, fire or burnt ground wherever they went. In addition to land management, fire and smoke are central to virtually every aspect of daily life, and to every life

---

6   Following this section in the manuscript, a heading, 'Restraint', shows that Debbie intended to discuss this concept, but never managed to do so—eds.

passage. Birth, initiations, dispute resolutions and funerals all require fire and smoke. Rights to use fire in particular contexts are allocated among kin and defended in the same way that rights to songs, designs and other forms of knowledge are defended. According to John Bradley:

> It is important to note that burning country is not just fire, smoke and blackened vegetation. Firing country involves people who have ways of interpreting their place within the environment where they live, on the country they call home. Their relationship with fire at its most basic is as a tool, but fire is also related to events associated with the past and the future, events which to the outsider may not be considered that important, but to the indigenous community are very important. Fire, then, can be seen to be a part of an ecology of internal relations; no event occurs which stands alone. An event such as the lighting of country is a synthesis of relationships to other events. Fire is but one event which is related to many others. (1995, 31)

Some general considerations

On a pragmatic level, fire-stick farming involves getting rid of long grass and grass seeds which impede travel. It means being able to see the animal tracks, and thus to hunt better. It means being able to see snakes and snake tracks so as to avoid them.

Burning benefits other animals: new growth is up to five times richer in nutrients than old growth (Braithwaite 1995). The benefit to others is also good for hunters. To quote April Bright, a traditional owner of floodplains in the area between the Finniss and Reynolds Rivers:

> 'Burn grass time' gives us good hunting. It brings animals such as wallabies, kangaroos and turkeys on the new fresh feed of green grasses and plants. But it does not only provide for us but also for animals, birds, reptiles and insects. After the 'burn' you will see hundreds of white cockatoos digging for grass roots. It's quite funny because they are no longer snow white but have blackened heads, and undercarriages black from the soot. The birds fly to the smoke to snatch up insects. Wallabies, kangaroos, bandicoots, birds, rats, mice, reptiles and insects all access these areas for food. If it wasn't burnt they would not be able to penetrate the dense and long spear grass and other grasses for these sources of food. (1995, 60)

Controlled burning has among its aims that it will not wreak havoc on animal life. East Arnhem Land/Yolngu man Joe Yunupingu explained:

> I care of the fire. The fire burnt only traditional way. Because we look after the animals, birds and land. The land is real important for us. Our lands. If we want to go make a fire, to burn, every year not to fire, every year. Take about two, three year for the right time got to be burnt. Got to look for animal. Kill animal, few, not much. We look after the animals, eat them not to waste it ... That's the law for the Yolngu people. (1995, 65–66)

In addition to fire to promote animal life and to facilitate hunting, there is also evidence of the use of fire to manage harvests. One method was to spread the harvest out through time by successively burning small areas (Bright 1995, 62). Another method is to produce a large harvest all at once to generate surpluses for ceremonial gatherings (Hallam 1987, 52–53; Lewis 2002, 50; Mulvaney 1987, 88–89).

The knowledgeable use of fire depends on detailed knowledge of soils, landforms, surface and underground water, and types of vegetation, as well as time of year, time of day and type of wind. Bright, in her detailed discussion of fire, speaks of big winds and slow winds, hot burns and cool burns. She notes that different combinations are appropriate for different landforms and different times of year (1995, 60–65).

In addition, there are always areas which are not burnt. In many parts of Australia the area around a sacred site is kept free of fire. Some of these areas serve the function of refuge for plant and animal species. Protection of particular areas requires, of course, careful burning in the vicinity. Other areas which are not burnt include many forested areas, particularly rainforests. These areas are vital resource areas for foods such as yams and medicines and are essential habitats for a range of other animals.

The right to burn is one of the rights and duties of ownership of Country, and the corresponding obligation is to refrain from burning other people's Country. This links with a key point in starting a fire, which is knowing where it will stop. Only with this knowledge is it possible to avoid burning places that should not be burnt for ecological, social or religious reasons (and these three factors may be connected in many instances). For example, Nanikiya Munungurritj, an Aboriginal traditional owner in East Arnhem Land, and a ranger with the Dhimurru Land Management organisation, spoke about burning his Country, saying that you sing the Country before you burn it. In your mind you see the fire, you know where it is going and you know where it will stop. Only then do you light the fire.[7]

The ubiquity of fire is counterbalanced by the fact that there is also good evidence for regional variation in fire regimes. In the northern flood plains, for example, burning is initiated towards the end of the wet season. The appearance of browned-off grass is the signal to start burning, and the appropriate place to start burning, for ecological, pragmatic and other reasons, is along the higher Country that borders the floodplains (Rose et al. 2002, 20).

The knowledge of when and where to start burning is complemented by the knowledge of where the fire will stop. This knowledge is built up out of intimately detailed knowledge of the specific grasses, soil type, landforms and plant communities. The duration of burning varies. People continue burning to clear off long grass for as long as there is evidence (in the form of green grass) that there is enough moisture to slow a fire down and enable regrowth. In the monsoon tropics where the rains are relatively predictable, the main burning months are April, May and June.[8] As Bright indicated, there is a specific season for burning off the long grass. She calls it 'burn grass time', and she notes the moral imperative to burn: 'It is part of our responsibility [to be] looking after our Country. If you don't look after Country, Country won't look after you' (1995, 59).

In contrast, Peter Latz (1995, 31, 34) reports that in the desert Aboriginal people burned throughout most of the year. Fire was an important tool in hunting, and the desert fires were by preference restricted in extent. Seasonal variation is extreme in these unpredictable environments, and rain is not strongly linked to annual cycles. Aboriginal people reportedly did not favour hot summer fires. There was, therefore, no annual regime, but there was avoidance during the hottest times of year when possible.

The use of fire to create a mosaic effect is well documented for spinifex Country and is also well documented for some of the landforms in the tropics (Latz 1995, 33–34). Recent research indicates that fire ecology

---

7   Personal communication with Nanikiya Munungurritj during the Bushfire '97 Conference, Darwin, 8–10 July 1997.
8   Information provided by Nancy Daiyi, Margaret Daiyi and Linda Payi Ford.

throughout Australia, from Tasmania in the south to the most northern regions was managed to create ecological mosaics (Latz 1995, 34; see also Gammage 2002; Jones 1969, 224–28; Stevenson 1985).

In sum, at a general level it is possible to say that there was a system of fire management that involved interruption of 'natural' fire regimes, protection of certain fire-sensitive species, use of fire regularly to control scrub, use of fire to time harvests of certain plants, use of fire to attract game, use of fire to produce a particular aesthetic that signals growth, fertility and clarity of vision, and use of fire to produce mosaics of micro-habitats. This latter point required knowledge of fire history as well as a range of habitats. While the specifics of how to achieve these ends shift across ecological zones, the purposes seem remarkably stable.

Into the Victoria River Country

There are several major ecological zones which are defined by Victoria River Aboriginal people today and which in pre-contact times would almost certainly have been managed under different fire regimes. These zones are the higher rainfall region north of the Stokes Range; the riverine savannas inland from the Stokes Range; and the dry watershed regions and mesa tops. It is most unlikely that there would have been uniformity of fire regimes within the Victoria River District.

Explorers and settlers

An examination of explorers' accounts of their travels in the Victoria River valley shows that they observed extensive, even 'luxurious', grass growth at all times of year. This is to say that they did not observe the effects of large-scale burning at any time of year, either north or south of the range.

They did note areas that had been burnt, and they were quick to observe that new grass grew readily and that kangaroos were attracted to it. James Wilson, a member of the North Australian Expedition, noted:

> In April and May, when the grass becomes dry, they burn it off about such water-holes and creeks as the kangaroos frequent; when the grass is thus early burned, the roots being still moist send up a second crop, and this is so sudden that I have seen green grass the third day after the dry was burned. This is done by the native to induce the kangaroos to come to such spots to feed and be the more convenient for him to hunt. (1885, 151)

During the entire expedition Wilson was based on the lower Victoria 'saltwater side'. His statement identifies April and May as 'burn grass time' and is thus consistent with the floodplain evidence discussed above. A compilation of explorers' observations shows that Aboriginal people were making fires in every month of the year, as one would expect (Lewis 2002, 54–61). It is rarely possible to specify the exact purpose of the smoke or burnt grass that was observed, and the general knowledge extant today suggests that any fire may have had multiple purposes.

When settlers came into the Victoria River Country in the early 1880s, they saw fires and they began to think that Aboriginal people were using fire as a weapon against them. This view was not unfounded, from either a white fellow or an Indigenous perspective. The explorer Augustus Gregory left some of the members of his party at a depot at Mt Sanford, while he proceeded inland on his explorations. On his return to the depot, he learned that his party:

> had been, however, somewhat annoyed by the blacks, who had made frequent attempts to burn the camp, and also the horses, by setting fire to the grass, and on some occasions had come to actual hostilities. (1884, 143)

Subsequently, settlers and police suspected that Aborigines were setting fire for the purpose of burning them and their beasts out. They took a merciless line on fire, suppressing Aboriginal burning wherever and whenever possible. Their efforts were aimed at protecting their stock, feed for their stock, fences, homes and lives. Among the losses were the habitats of animals who depended on fire, the mosaic of habitats produced by Aboriginal burning, the diminution of distribution of fire-dependent species, the loss of balance between controlled and uncontrolled fires, and the opening up of land to woody shrub and other floral invasions (Chapter 8).

Settlers' repression of Indigenous use of fire was savage. It remains scarred into the consciousness of many of the older people I spoke with. The effect of their efforts is that in many areas very little detailed knowledge appears to have survived concerning the use of fire outside of camp and ritual contexts. Dora Jilpngarri, the oldest person now alive at Yarralin, explained when asked about Indigenous use of fire in the Country: 'No, they never used to burn. They weren't allowed to. Policeman would kill them, or manager would kill them. They weren't allowed to burn. They never used to burn.'

A number of people of her age cohort have asserted that Aboriginal people never burnt the Country, even before white fellows came. Jilpngarri's words show the ambivalence of the position. On the one hand she asserts that they never did burn. On the other hand, that people were not allowed to burn. The fact that burning was prohibited indicates that burning was being done, and that it was being brutally suppressed. As I will discuss shortly, the term burning has developed a context-dependent polysemy which goes some way toward clarifying the ambiguity of Dora's statement.

In both northern and southern regions, pastoralists also instituted burning regimes. Where Indigenous and white pastoralist burning regimes are similar, it is not possible to determine whether the pastoralists adopted Indigenous regimes, or whether Indigenous people adopted pastoralist regimes. It is possible, of course, that both groups were burning in response to the ecological demands of Country and independently arrived at similar regimes. However, the evidence shows Aboriginal people making fires for a great variety of reasons and making them throughout the year. The suppression of Indigenous fires targeted all types of fire, and it was also resisted in some instances. Daly Pulkara made this point in relation to signal fires: 'Fires are for signal—*kartiya* [white fellows] can't stop them. They put fire there for signal.'

My working hypothesis is that prior to white settlement, Aboriginal people in this district used fire in ways that are consistent with the general system discussed above. The further hypothesis is that people would have developed fire regimes suitable to their savanna homelands and that there was a distinctive savanna burning regime, with its own seasonal indicators and its own aesthetic. The hypotheses cannot be proved beyond doubt, but the information is, in my view, sufficient to confirm that these are reasonable hypotheses that merit further investigation in other savanna regions.

Daily life

The following purposes of fire are significant in the Victoria River District:

- for cooking;
- for boiling water;
- for warmth;
- for light;
- in ceremony (mortuary and other rituals);

- for knocking down dead trees for firewood;
- for cleaning up an area prior to camping;
- for healing—to create warmth and steam, using medicinal plants (also gender-restricted rituals);
- to make the ashes used with chewing tobacco;
- as part of the process of leaching toxins out of certain foods to make them edible;
- to drive away dangerous supernatural figures;
- to erase the traces of life so that dead people will not want to return;
- for hardening spear points and digging stick points;
- to anneal stone to make it better for working into tools;
- to alter the chemical structure of haematite, transforming yellow ochre to red ochre;
- for communication—signalling peoples' presence in an area;
- in hunting (no longer allowed):
  - to attract animals to a place where they can conveniently be caught;
  - in conjunction with hawk-hunting hides;
- as a system of land management (fine-grained detail is lacking, but the concept appears to have been present, although the specific practices are no longer allowed).

Fire for hunting

The main uses of fire in hunting are in clearing out old grass to allow for green pick, and the use of fire in hawk-hunting hides. The first is consistent with Indigenous practices throughout the continent; fires that burnt off dry grass at a time when new grass would grow facilitated hunting by bringing animals close to the hunters.

The second is a technique that is unique to an area largely coterminous with the Victoria River District. In this region Aboriginal people used to build small stone enclosures with roofs of branches and grass. One or two men would sit in the hide while others set the surrounding bush on fire. Using a small bird as bait on the roof of the hide (spinifex pigeon was identified as preferred bait), the hunters in the hide enticed the circling hawks to dive for the bait, and then grabbed them and broke their necks (Lewis 1988[9]).

---

9   Lewis also discusses some of the comparative literature from other parts of the continent.

Timing indicators

Many of the uses of fire mentioned above are not restricted to any time of year and would have been used throughout the year. Fires for cooking, medication, cleaning up a camp, signalling and other daily activities were omnipresent. Fires in funeral ceremonies were of course dependent on the timing of death, but had no particular seasonal component, as far as is known. Fires in other ceremonies were linked to the ceremonial cycle. That cycle has changed with white settlement. According to informants, ceremony formerly took place after the wet season when resources were rich enough to allow large numbers of people to gather and stay together. After the pastoral regime was established, ceremonies were rescheduled to take place *during* the wet season, as this was the period when Aboriginal people were released from station work and allowed (indeed required) to go bush.

Aboriginal people south of the Stokes Range were adamant that before white fellows arrived in the Country, burning was initiated after the first or second rains, at a time when green grass was appearing. This is consistent with the pastoral regime south of the ranges, where firing began at this time. This information was offered by a number of people independently, and on separate occasions spanning 20 years of research. Perhaps the most interesting account of burning was offered by an old man, now deceased, who explained that the time to clean up or clear up Country is when new green growth appears. The presence of green growth as an indicator for burning is an extremely context-sensitive indicator. It first appears with the first rains and continues throughout the wet season and into the dry. The emphasis, however, is on first appearance, as the time to start burning. It thus suggests a cycle of fire that begins with the early rains, and probably continues whenever possible through the wet season.

This contrasts with the floodplain data, where the indicator for initiating burning is the appearance of dry-brown grass. The contrast reflects the different rainfall regimes (1,514.9 mm annual mean rainfall, Finniss River; 632.6 mm annual mean rainfall for VRD station). Different rainfall zones produce different problematics and different conceptions of the annual cycle. High rainfall requires people to remove grass, and people develop an annual cycle that begins with the appearance of dry grass after the rains. Low rainfall requires people to curate new grass, and people develop an annual cycle that begins with the appearance of new grass when the rains start.

The green grass indicator is specific to any given place, soil type and plant community, and thus has the quality of localised knowledge and practice that is the hallmark of Aboriginal knowledge. It contrasts with the pastoralist regime. According to the few Aboriginal people who spoke of pastoralists' burning, after the first or second rain the manager would send out some young Indigenous men on horses with matches, telling them to burn the old grass (see Lewis 2002, 63–70). That regime apparently afforded little sensitivity to the localisation of rains, the rates of regeneration of different plant communities and the differential effects of fire on different plant and soil types.

The idea that the imminent approach of the wet (or the appearance of new green growth) is the appropriate time to initiate burning of old growth is consistent with information collected from Aboriginal people in other parts of the semi-arid savannas.[10] Fiona Walsh and colleagues (2003) have recently queried this finding, suggesting that it is a borrowing from pastoralists and that before settlement people would have burned in the cold season. The people interviewed at Yarralin and Lingara (south of the Stokes Range) insisted that they did not burn in the cold weather.

North of the range in the higher rainfall saltwater side zone, there would almost certainly have been more intensive burning, more burning at the end or beginning of the dry season, and perhaps also more attention to the need to know where fires were going to stop. In this spear grass Country, there was a clear imperative, shared by Aborigines and pastoralists alike, to render the country accessible by removing excessive growth. It is probable that the Indigenous burning regime was more like that described for the northern floodplains and other spear grass Country. It may be significant that the most detailed information available on the social management of fire (discussed below) was obtained north of the Stokes Range.

Pragmatics

South of the ranges, people were adamant their ancestors never burnt large areas at once. The explorers' accounts bear this out. People spoke of the usefulness of patches of tall grass behind which a hunter would hide when stalking. They also spoke of the fact that kangaroos and other animals eat

---

10   Kimber and Smith (1987, 221–23); see also Kimber (1983).

grass. They regarded extensive burning as both wasteful and dangerous. It was wasteful in destroying grass that was needed by both herbivores and hunters. The concept of danger will be discussed below.

North of the ranges, as discussed above, the pragmatics of dealing with spear grass provided strong reasons for burning off after the wet season.

There was general agreement south of the Stokes Range that people should burn in their own Country, and not burn in other people's Country. There was general condemnation of fires that get out of control (from the station rubbish tip, from tourists, from kids 'mucking around' and other sources). It is probable that people's right to burn in their own Country was matched by a corresponding prohibition on burning other people's Country.

North of the Stokes Range, in contrast, people stated forcefully that to set a fire that burned into someone else's Country was to commit a capital offence. The punishment was death, at least in principle.

Aesthetics

Lesley Head (1994) has carried out extensive research into fire in the Kimberley, and her findings (in a higher rainfall zone) suggest that Aboriginal people have an aesthetic of fire that values the look of burnt Country. She hypothesises that this fire regime, and thus the aesthetic, is ancient (several millennia at the least), and that it is ongoing. Walsh and colleagues (2003) have contested this analysis. In their view, the aesthetic of burnt Country is recent, and represents Aboriginal people's response to returning to Country that had not been burnt for a long time. In burning again, people restored Country to a state that was visually pleasing. Walsh et al. indicate that they do not accept that this aesthetic formed a part of pre–white fellow burning.

It is not possible to resolve this difference of opinion for all times and places, and it is best to consider the strong possibility of regional variation. It seems certain, however, that Head is correct in her assessment of the long-term continuity of an aesthetic of burnt Country for some regions. A crucial piece of evidence comes from Bathurst and Melville Island, where Andrée Grau (2005) has for many years studied the aesthetics of dance. She has found that Tiwi dance calls for clean, clear bodily motion and clearly articulated bodily shapes. Tiwi people consciously link the aesthetics of dance with the aesthetics of burnt Country. In both dance and Country, the desired state is one in which clear clean lines are visible, in which angles are demarcated and in which shapes have strong sharp outlines. It is almost inconceivable that this strong association of dance and burnt Country could have arisen within the recent period of white contact. The logical probability is that this is an old and deeply internalised aesthetic.

In the Victoria River District, people south of the ranges articulated an aesthetic of 'clear' Country. Their position was that Country looks good when you can see. Keeping the grass and scrub from obscuring vision, and keeping the trees from becoming too dense, were both articulated as ideals for a clear Country. This aesthetic of clarity has pragmatic implications for hunting, of course. Country where the hunter can see is Country in which the hunter can hunt. Alternatively, the aesthetic of clarity was balanced by a concern that Country not be denuded. The logic here was that hunters need to have something to hide behind, and so do animals. The balance between clear Country and denuded Country suggests that in this region, as elsewhere, a fine-grained mosaic would provide the balance being discussed.

North of the Ranges, the pragmatics and aesthetics were almost certainly generally consistent with those reported for well-watered regions. They include the concept of clearing or cleaning excessive growth so that people and others can move around. Clarity of vision was also part of this aesthetic.

In contrast to the Tiwi aesthetic of clean, articulated shapes, Yarralin people express a savanna aesthetic. Clarity of vision across distance is the heart of the aesthetic. Jessie Wirrpa expressed her sense of this aesthetic in another context. Years ago, she was taken to Adelaide for an operation. Describing her time there she said that she was unhappy because she couldn't see anything. There were too many buildings in the way.

Riley Young spoke to the loss of this aesthetic in his own home Country where the woody weed invasion (Chapter 8) is destroying clarity of vision. He and others defined the loss of clarity as *marnin*—glossed as 'you can't look', or 'shut im up, can't get through'. Both glosses vividly communicate the sense of trying to see, hunt and travel in scrub-invaded Country.

An ethic of care

The people interviewed in this study articulated a strong view of their use of fire, consistent with the culture of fire articulated in myth. That is, they claim for themselves and their forebears the ability to use fire carefully and productively. Allan Young was eloquent on this point: 'They never burned the Country. They bin holdem Country. They bin holdem Country.' The term 'hold' is Aboriginal English; it bears connotations of sustaining. The idea of holding Country indicates the human responsibility to interact with Country in ways that sustain it in a condition that continues to benefit the creatures, including human creatures, who live there (see Chapter 9).

These assertions clearly gain rhetorical force in the context of pastoralists' broadacre burning and the out-of-control fires that are now prevalent. However, it is not the case that these views have arisen only in response to white fellow burning. They are consistent with views that are so widespread, and so prevalent in areas where white fellow influence has been minimal, that they clearly articulate an Indigenous ethic of care that is given added emphasis because of the contemporary fire situation.

My teachers were particularly vehement about not allowing fires to get into sacred sites. Their views thus rely on the implicit assumption that fires lit responsibly would be lit by people who know the Country and who know fire. That is, fires were and are lit by people who know where the sacred sites are, what the terrain is like and other factors influencing fire. They know how the fire would behave—where it would go, and where it would stop. 'Holding' the Country, and preventing damage, thus requires the local, detailed knowledge of ecological and sacred geographies within which the lighting of domesticated fires is embedded.

Wild and domesticated

The evidence from mythology speaks to a culture of fire. Two main aspects are noteworthy. The first is fire as an out-of-control source of danger. In myth, these fires race through Country after Country, crossing social boundaries and burning Country and animals with harmful effects. Sometimes these events are configured as acts of aggression, and in other stories they figure as events that just happen. These stories define out-of-control fires as extremely negative events. Today, in people's daily discussions of the out-of-control fires that ravage the region, the harm is identified as direct harm to humans, animals and Country and also as potential harm arising from Dreaming. If fires burn out sacred areas, damage to the sacred site of an area produces negative effects for the traditional owners and, in some instances, for people and other living things as well.

The second aspect of fire is that it is a central feature of human life. The centrality of fire to humanity is symbolised by the fire-stick. In myth, fire-sticks are associated with women, and thus with kinship, camp, cooked food and other major signs of human life as distinguished from the lives of animals.[11] Fire-sticks are associated with ceremony and with cleaning up areas for camping. They are never associated with the large out-of-control fires discussed above.

Evidence from myth thus indicates a distinction between wild and domestic. We have encountered this distinction before in relation to dingoes. In the context of fire, it prompts the view that Aboriginal people in this region regard their own use of fire as a practice comparable to what in other parts of the world is called domestication. They have taken a wild element of the world and brought it under control by human effort and knowledge. In contrast to most parts of the world where humans domesticated plants and animals, Australian Aborigines have domesticated fire.

The distinction between wild (out-of-control fire) and domestic (fire that 'holds' the Country) has acquired new meanings in recent years. People now contrast their own fires with the 'wild fires' that rage across the country, started accidentally by tourists, or breaking out of rubbish tips, or getting away from rangers and pastoralists who are trying to burn responsibly but miss the mark.[12]

---

11 Gender-restricted myth and ceremony enhances these understandings.
12 A note at the end of this chapter shows that at this point Debbie wanted to discuss 'a nomadics of fire'—eds.

8

Wild and Ugly

In 1986 Daly Pulkara and I were travelling from Yarralin to Lingara. The route was familiar to us both, and we stopped partway because I wanted to video some of the most spectacular erosion in the Victoria River District. I asked Daly what he called this Country. He looked at it long and heavily before he said: 'It's the wild. Just the wild.' Daly went on to speak of quiet Country—the Country in which the care of generations of people is evident to those who know how to see it. Quiet Country stands in contrast to the wild: we were looking at a wilderness, man-made and cattle-made. The life of the Country was falling down into the gullies and washing away with the rains.

Wild Country is what Hobbles would have described as 'disorganised' in contrast to Indigenous people's organisation. It includes the erosion gullies, washaways, scald areas, zones of noxious weed invasions and zones of woody weed invasions. It also includes the eco-places of loss: where certain plants used to grow, where billabongs used to hold water and be filled with lilies, where Dreaming trees stood, and native animals used to be found in abundance. In the pastoral country of North Australia, face-to-face encounters with the wild are impossible to avoid. However, it takes the knowledgeable attention of those who belong there to know how devastating are the absences in these wild places.

Figure 8.1. Eroded Country on Humbert River Station that Daly Pulkara described as 'the wild, just the wild', 1981.

Source: Photograph by Darrell Lewis.

I have discussed concepts of change that do not produce 'the wild' (Chapter 5). In this chapter I examine contemporary changes in the land primarily through the perspectives of my Aboriginal teachers. My purpose is to draw out the ramifying effects of damage in order to gain a stronger sense of the losses entailed by double death. The idea that environmental change is somehow separate from social and cultural change is of course erroneous. Few would argue that there are not 'social impacts'. My point goes deeper: these are not impacts of one order on another, nor are they linear, as the term 'impact' seems to imply. There is just one order here; it is the experience of life within a ramifying and increasingly recursive field of devastation.

Irreversible change

The history of the Victoria River District has been extremely brutal, a fact of which Aboriginal people are acutely aware (Rose 1991). In analysing their own history, they identify two major moments in the colonisation process. I have labelled them invasion and settlement. While the one necessarily precedes the other, Yarralin people generally speak of them not as periods but as processes. Invasion was the process by which Europeans came with cattle and guns, and with the intention to settle. According to Victoria River historians, and here I draw particularly on the work of Hobbles Danaiyarri, the Europeans' strategy was first to kill people and then gain control of the land and the surviving people. Some of the old people with whom I studied spoke with bafflement over the fact that white people thought the lives of cattle to be of greater value than the lives of Indigenous people. Others were bitter that white people had used Aboriginal labour for the really hard jobs in preference to the labour of animals, because they did not want to wear out their animals.

Settler Australians' culture of cattle included the understanding of cattle as a special kind of living property. The word cattle has the same etymological root as capital and as chattel. The term 'goods and chattels' used to be 'goods and cattels'. Long before the invention of money, cattle were one of the first forms of moveable wealth (Rifkin 1992, 28), and for the pastoralists on the frontier of North Australia, cattle were densely significant in terms of property, wealth, livelihood and culture. For settlers, land was a condition for wealth, and a commodity for transactions which might produce wealth, but there is not much evidence to suggest that the settlers actually understood land to be a source of wealth in itself. They set about wasting it extravagantly. Their initial actions were to clear out the Aboriginal people as much as practicable, and to suppress their use and care of the Country.

The Northern Territory newspapers for the 1880s, 1890s and 1900s are full of headlines about Aboriginal people murdering whites. They give the impression of a type of guerrilla warfare in which blacks made unprovoked attacks and whites retaliated. White people's view of Aborigines as primitive (at best) and as obstacles to settlement ensured that they would not recognise the knowledge and the practices of care with which people curated and sustained the Country. Accordingly, people, including their knowledge and practices, were also wasted. Cattle came first. On a Territory-wide basis enough murder took place to supply a string of headlines that generate a sense of the battleground, but on a local basis the story is quite different. Reading the newspaper accounts and comparing them with the local fine-grained detail of the police accounts one learns that most of the patrols and most of the killings were in response to disputes or alleged disputes about cattle. Constable William Willshire, the first policeman along the Victoria River, published his experiences in a book, as well as keeping his official journal. He describes the most terrible bloodshed, and most of his accounts of what clearly are massacres begin with a casual remark about going in search of cattle killers or coming across the tracks of some cattle killers. One example will suffice to indicate the link between cattle and killing.

> In the month of June, 1894, we came across some tracks of natives that had been recently killing cattle on the Victoria Run. We followed them along ... Next morning we picked up the tracks and crossed the river and in two hours we came upon the cattle killers camped close to the river ... They commenced running and many of them escaped in the tropical growth ... Next morning we went on, picked up another set of tracks ... and came upon a large mob of natives camped amongst rocks. (Willshire 1896, 40–41)

In this account it is never made clear how he knew that the tracks he followed had been made by people who had been killing cattle, and this problem besets him throughout the whole of his appointment—he found tracks and followed them in the expectation that they must be the people he was looking for (see Rose 1991, Chapters 3 and 9). This expedition led him to horrible action. He wrote:

> They scattered in all directions, setting fire to the grass on each side of us, throwing occasional spears, and yelling at us. It's no use mincing matters—the Martini-Henry carbines at this critical moment were talking English in the silent majesty of these great eternal rocks. The mountain was swathed in a regal robe of fiery grandeur, and its ominous roar was close upon us. The weird, awful beauty of the scene held us spellbound for a few seconds. (Willshire 1896, 41)

Willshire had tracked the 'killers' from close to the Victoria River Downs (VRD) Centre Camp, across Country and into the edge of the Bilinara sandstone known as Pilimatjaru, where surviving Bilinara people took refuge. He kept coming upon more tracks which he claimed were those of cattle killers and, in the end, he seemed to be killing people who were not necessarily even the same people as the first mob he was tracking. The whole thing rests on an implicit assumption that it is proper to kill people who have been or who might be thought to be planning to interfere with cattle.

As Willshire's account shows, in the early years Aborigines had used fire as one of their weapons in the war of survival. The explorer Augustus Gregory left some of the members of his party at a depot at Mt Sanford, while he proceeded inland on his explorations. As earlier noted, on his return to the depot he learned that Aborigines had tried to burn the camp and the horses by setting fire to the grass (1884, 143).

Subsequently, settlers and police suspected that Aborigines were setting fire for the purpose of burning them and their beasts out. They took a merciless line on fire, suppressing Aboriginal burning wherever and whenever possible. Their efforts were aimed at protecting their stock, fences, homes and lives. Among the losses were the habitats of animals who depended on fire, the mosaic of habitats produced by Aboriginal burning (Chapter 7), the diminution of the distribution of fire-dependent species, the loss of balance between controlled and uncontrolled fires, the opening up of land to woody shrub and other floral invasions, and the loss of much of Aboriginal people's detailed knowledge of the use of fire to sustain the Country.

~ ~ ~

The first settlers saw environmental change almost from the start. With the arrival of heavy, hard-hoofed cattle, the riverbanks were cut up and erosion began. Within a few years the riverbanks began to slip into the rivers, and the soils became compacted. Native grasses and forbs were either eaten out or lost because of soil changes; many of them were the preferential food for cattle, and the stocking and overstocking, along with changes to the soils, meant that these grasses and forbs were quickly reduced. Many of these grasses produced seeds that were staple foods for Aboriginal people.

Reports of erosion continued on a regular basis. Drought years worsened conditions, years of plenty enabled the Country to come back to some degree. Cattle had to have water, so many of their greatest impacts were along the waterways, while on the back blocks away from the rivers, cattle ran wild, along with the wild horses, donkeys, camels and occasional water buffalo (Lewis 2002).

In 1945 the geographer Wilson Maze estimated that 4 to 12 inches of topsoil had been lost in the Ord River area (adjacent to, and subjected to similar pressures as, the Victoria River); he warned that soils and plants subjected to such pressure could not sustain an industry (Maze 1945, 7–19). His voice was ignored. In 1969 a survey showed that 'twenty per cent of the entire Victoria River District was suffering from accelerated soil erosion' (Letts). The survey was announced by Dr Goff Letts, Director of Primary Industry, who was introducing a Soils Conservation and Land Utilisation Act into the NT Legislative Council. A decade or so later, the Conservation Commission of the Northern Territory warned that 'signs of deterioration through fertility loss and erosion is [sic] already evident' (Melville 1981, vi). Ian Melville advocated the use of 'improved' (meaning introduced) pastures as one way of combatting the overgrazing that leads to erosion.

A study of global environmental crises identifies a number of types of land degradation that are prevalent in New South Wales (where the situation is relatively well documented), and are likely to be present, and probably prevalent to greater or lesser degrees, in the Victoria River District: soil erosion (particularly wind and water erosion), soil acidity, soil structural decline, woody shrub infestation, lack of tree regeneration, landslides and salinity (Aplin et al. 1996, 50). Of these, erosion and woody shrub infestation are rapidly increasing in the Victoria River District (Lewis 2002). Water degradation is another major problem: groundwater depletion and pollution are key problems (Aplin et al. 1996, 58). The former premier of the state of Victoria offered a harsh but succinct summary: 'We could not have made a bigger mess of the soil of this country than if its destruction had been carried out under supervision' (quoted in Beale and Fray 1990, 121).

Melville discusses accelerated forms of soil erosion caused by factors that include clearing, ploughing, grazing, off-road vehicles, mining and other forms of human interventions. In the monsoonal tropics, factors which remove the vegetation are particularly significant because the earth becomes so dry during the dry season and is then exceedingly vulnerable to the heavy downpours of the early wet season. In areas where there is heaving stocking, that is around watering points and in holding paddocks, soil is carried away by the wind. The removal of topsoil of course makes regeneration 'difficult or impossible' (Melville 1981, 5). Permanently bare areas are called 'scalds'; they continue to erode by the action of rain and wind. Trees, too, are affected. Grasses and herbs shade and cool the soil surface; with their loss trees find it difficult to cope with soil desiccation and they can die. This is one phenomenon that is termed 'desertification' (Melville 1981, 5).

Wind erosion also destroys vegetation through a sand blasting effect. Some of the big winds of the late dry season carry topsoils and sand; they drive the dust along like a huge red war engine. In contrast, rainfall erosion varies with the intensity of the falling rain. When rain falls on bare soils it smashes soil aggregates and blasts them into the air. The higher rates of rainfall in the tropics have a harsher erosive impact than the more moderate rains of the temperate zones (Melville 1981, 10). Soil particles removed in resulting flows form small channels or rills. Larger rills become gullies. When the water hits the rivers, the soil load enhances riverine forces to produce slumping and undercutting (13). Rivers widen, trees are washed away, banks are undercut and start to collapse, more soils flow into the rivers and the processes continue. Harry Recher, one of Australia's leading ecologists, describes water degradation as the 'great unseen, unspoken and unrecognised threat to our survival' (quoted in Beale and Fray 1990, 48).

Figure 8.2. A scald area in the Victoria River Country, Camfield Station, c. 1990.

Source: Photograph by Darrell Lewis.

Figure 8.3. Trees killed by overgrazing and drought, Wave Hill Station, c. 1988.

Source: Photograph by Darrell Lewis.

For most of the period of pastoralism in North Australia, government action has been directed toward supporting cattle as an industry rather than supporting the resource base on which it depends. North Australian pastoralism is sacrosanct. The relationship between the industry and the state is extremely close, and so-called threats to the industry are taken very seriously; indeed, they are spoken of as if they were akin to treason. Anything that appears to be a critique of the pastoral industry, and of development more generally, is likely to elicit an emotionally charged defensive response.

The economics of the pastoral industry differ from the rhetoric of the Northern Territory Government's promotion of the industry. No one would deny that fortunes have been made in the cattle industry, but not all pastoral properties have rewarded their owners handsomely. Jim Rawling summarises a number of issues:

> While important tax concessions are available to rural enterprises in Australia generally, and to private and public companies in particular, the NT historically has had further tax benefits over and above other Australian states. Additional benefits accrue to larger pastoral companies which have preferential access to emerging international project financiers. (1987, 31)

In 1967, for example, Rawling (1987, 20) found that pastoral rentals in the NT netted the Commonwealth $139,802.00; expenditure on infrastructure for watering and transport came to $4,119,291.00. Tax concessions, tax loss farming and land speculation were and remain part of the pastoral industry.

> International venture capital has had a particularly significant role in the purchase of NT pastoral holdings and can be seen to have strategies which have little to do with national interests, however defined. The outright profitability of pastoral activities within large corporations has not always been necessary as part of their land holding strategies. Interests which are little concerned with the viability of beef production can invest in rural leaseholds for several reasons. (Rawling 1987, 29)

In addition to the contradictory quality of the rhetoric of production when measured against the economics of production in many instances, there are also ecological issues. John Holmes's (1990) study of pastoral properties in the Gulf region shows that the grid of pastoral properties overlies a range of ecosystems, some of which are suitable for intensive pastoral activity, some of which are suited for supporting or intermittent pastoral activity, and some of which are not suited to any sort of pastoral activity at all. A few stations are unsuitable for pastoralism in their entirety, but most encompass a mix of ecosystems, and for the most part they encompass ecosystems in various stages of decline.

In a rational world these boundaries would be redrawn to enable a more productive conformity between ecology and economy. This is not a rational world, however, and there is still money and political capital to be gained from the cattle industry. Pastoralists had long maintained that the nature of the pastoral lease interfered with their ability to make long-term management decisions. Leases were allocated for long periods of time (100 years) and were subject to numerous covenants: bores to be maintained, weeds to be eradicated, feral animals to be kept under control; continued access for Aboriginal people to the natural waters and native animals of the lease area; stocking levels to be kept above certain minima, and more.[1]

In the Victoria River District, there is not much popular support for the public expression of a sense of crisis. Even well-grounded concerns about the future of the industry tend to shy away from the embedded problems of collapsing ecosystems. An expert who preferred to remain anonymous put it to me that the Northern Territory's Conservation Commission could only afford to repair about 1 per cent of the damaged Country, and that there is no way that pastoralists would be able to afford repairs, assuming that they would want to (and of course some do). In 1996 the Commonwealth Government committed $14 million over a period of six years toward the establishment of the Cooperative Research Centre for the Sustainable Development of Tropical Savannas (CRC-TS). The emphasis was not wholly on the pastoral industry, but pastoralism is the dominant industry in the tropical savannas and accordingly it received a great deal of research attention from scientists across a range of research institutions. At the end of the period of the CRC, the question remains: is pastoralism a viable industry? If so, what conditions would ensure long-term viability?

One of the ironies in the recent history of pastoralism in the Northern Territory is that a number of cattle stations have been purchased for Aboriginal people and have been the subject of claims. In those land claims, the Northern Territory Government opposed the claims (until about 1993) supposedly on the grounds that Aboriginal people would not be able to run the stations properly, and that the Territory's economic future was being

---

1   In the manuscript Debbie had a note here: 'fill in on leases, CSIRO recommendations, freehold' —eds.

undermined by Aboriginal people. I call this an irony, because a great deal of evidence suggests that if Country is left to recover, and if people work to manage fire and other ecological 'tools' so as to assist Country to recover, it is possible for Country to 'come back', at least to some extent. If the government was correct, and Aboriginal traditional owners let Country become unproductive from a pastoralist point of view, their actions would probably work toward saving the Country. In addition, one would want to note that many of the stations purchased for Aboriginal people and claimed by them were marginal, and hence affordable. One could argue, in fact, that funds for the purchase of stations for Aboriginal people were bailing out the cattle industry. A further factor is that almost all Aboriginal claimants on cattle stations wanted to run cattle. Their aspirations were not, as the government suggested, to drop out of the pastoral industry, but rather to enter it on their own terms. As yet there have not been studies to show whether Aboriginal methods of running cattle impact differently on ecosystems.

Bang, bang

Daly Pulkara contrasted Aboriginal ways of doing things with white fellow ways, noticing, in particular the issue of waste:

> I reckon you [kartiya], you're wasting. Bang, bang, everybody start from anywhere. Make im frightened, yeah. [But us, we] Just walking got a spear, and you can see something [an animal] there quiet, and people can sneak up and just have a look at that thing.[2]

Daly's words validate not only a method of hunting, but a method of footwalk knowledge that is built up out of, and finds its expression in, attention. He and others hunt with rifles too. The point of his comment is not to claim some sort of purity, but rather to assert the specific values of his way of doing things. Aborigines are often accused of wasting things, and Daly wanted to turn these words and point them in a different direction. His way of doing things is characteristic of his practice and is part of a way of knowledge that he has inherited from his forebears and has added to in the course of his own life. It is a way of knowledge that is attentive to the world.

---

2   Daly Pulkara, tape 80, recorded at Lingara, 15 July 1986.

Daly spoke to a white perception that Aborigines waste things, and he wanted to turn those words around and make a case that white people waste things by being indiscriminate and unobservant. Anzac Munnganyi, from Pigeon Hole, said: 'White people just came up blind, bumping into everything. And put the flag, put the flag.' His imagery of white people stumbling around in unknown country and yet having the arrogance to 'put the flag' and claim the land strikes me as immensely insightful. For settlers and for many Aboriginal pastoralists, pastoralism is a form of production. In the remainder of this chapter, I examine its underside as a form of destruction.

I have already told about going for yams and lilies, and ending up at the river fishing when the yams weren't there, or the billabong was bare. Individual episodes were usually treated as aberrations, and my teachers held to the view that if we went to the right place at the right time with the right people we would find what we had been looking for. Our 'bad luck', it seemed, was simply that.

Once I began making a concerted effort to document plants I asked more persistently, and some of my teachers became determined to find a particular plant and show it to me. We became far more methodical in our efforts, both in discussing where, when and how we would find a particular plant, and then setting out to do so. Once such plant is the toxic tuber *kayalarin,* which is emblematic of Jessie Wirrpa's home Country. I learned a lot about the tucker: what it would look like if we could see it, how we would gather it, and how we would process it. I think that it may have been *Crinum augustifolium,* as Ian Crawford (1982, 40–41) describes a toxic tuber in the Kimberley that grows in a similar habitat and is processed in the same way, and with a similar appearance to *kayalarin.* We were never able to find a single specimen. It used to grow in such profusion that it became an identifier both of home (for the people) and of the people themselves (to others). One of the ancestors was named for it—Old Kayalarin; the area where it grew in profusion is also called by its name—Kayalarin Country. We drove and walked all over the area where it used to be, and there was no *kayalarin.* We fenced off a small exclosure in the area where it always used to grow and checked it over several years to see if anything grew back when the cattle and horses were kept out. Jessie experienced our defeat quite personally, becoming increasingly depressed every time we talked about *kayalarin.* After a few years we did not talk about it anymore.

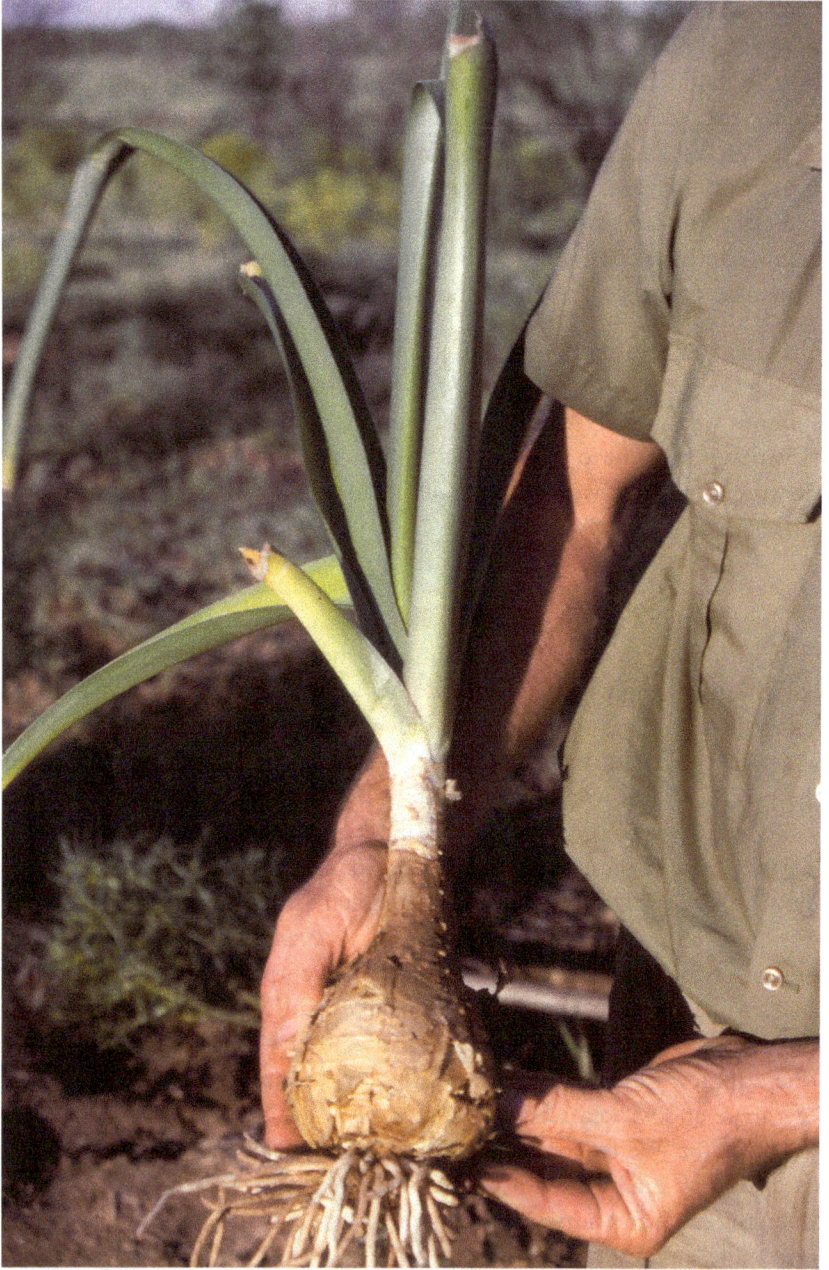

Figure 8.4. *Kayalarin* (*Crinum augustifolium*), a species once prolific on VRD, but now extremely rare.

Source: Photograph by Darrell Lewis.

The research prompted people to think about patterns of loss, rather than holding each event to be an aberration in a pattern of plenty. This was difficult, not only because it was disheartening, but also because it contradicted much that people hold dear about their relationships to Country and the ancestors. In *Dingo Makes Us Human* I explain in greater detail that a basic moral principle of Yarralin people's relationships to Country, and for them, a moral principle that is foundational to how the world works, is that a Country and its people take care of their own (Rose 1992, 107). People's responsibilities to place are reciprocated in the responsibilities that Country has to nurture its people. Our attempts to document the foods that nurture ended up in many instances documenting patterned and continuing loss. I discussed with Jessie the idea that if we could get a *kayalarin* plant she might to try to re-establish them. She said no; let them grow them in Darwin where people can take care of them, she said. They're gone here. Those plants were an identifier of her group and home. Other people's *kayalarin* were, apparently, no substitute, and nor did she want to subject plants to further decimation.

These are deep and serious issues about the quality of change and the continuity of life in the world. I will return to them, but first I want to document more of the loss. As the research proceeded, people began to tally up some of the areas and kinds of damage, and to try out explanations. Doug Campbell mused about loss. He spoke particularly of areas close to the station where people's sedentarised condition almost certainly led to overuse. As he points out, however, cattle were competitors for Aboriginal people's food, and they destroyed large areas.

Figure 8.5. Doug Campbell, Yarralin, 1981.
Source: Photograph by Darrell Lewis.

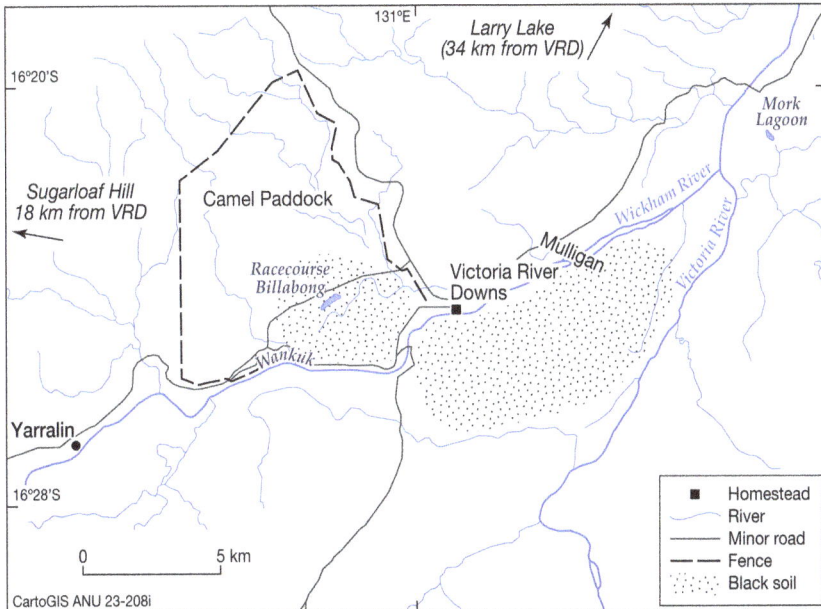

Map 8.1. The foraging range of Doug Campbell and other Aboriginal people who once lived in the 'compound' at VRD homestead.

Source: Karina Pelling of CartoGIS ANU.

It's not like before, everything used to grow every way. Camel paddock, too. And Racecourse billabong, coming out to the airstrip and going back to Wangkuk: there used to be two yams: *kamara* [*Ipomoea aquatica*] and *wayita*. I don't know where that tucker, might be too much cattle. New bore there, *wayita* was growing there, too, long time ago.

Might be some on that hill, there, Japarta [Mululu Bob] reckons *wayita* growing there. And *wanimirra* too. Rubber bush and bindii [introduced weeds, *Calotropis procura* and *Tribulus terrestris*] killing these yams too, because *wayita* used to be all around Sugarloaf [hill, near Yarralin].

At that black soil over at Little Mulligan, we no more see that [yams].

And *kitpan* [*Cucumis,* probably *melo*], should be on the black soil, all around, that's gone too. [From] Larry Lake and all around [black soil Country]. The cattle really like that, it's a little cucumber.

263

> Before, you'd walk all over them, and now you can't find them. All the tucker still coming [up] in Jasper Gorge—*kakawuli,* and *mamunya* [probably *Dioscorea bulbifera*]. And at Slatey too, not much cattle there.
>
> At Mutpurani, Mother used to get it there—it used to be the biggest garden.
>
> Even bamboo spear, that river was full, right up to Layit junction. Nothing there now. You can just see rubber bush and all the rubbish grass.
>
> No *kamara* at Larry lake and Mork billabong. And *karil* [another *Cucumis* sp.], we used to get plenty, and leave some there for next time.[3]

Dora also spoke of some of the plants that used to grow around Racecourse and Mulligan billabongs.

> And nother one tucker *kamara* [yam]. *Kamara* like a parsley, eh? Big one, like that. Get im like that, that *kamara. Wayita* all right. You know *wayita,* little one.
>
> Debbie: Was that *kamara* growing close up to VRD?
>
> Dora: Yeah. Im bin growing there before, longa VRD longa that old racecourse, this side, eh. Man call im *Kankiji* [billabong]. There now lotta *kamara* bin get up there before. When I was big one now. We bin havem big mob there.
>
> And *janaka. Janaka* bin longa that bull paddock there longa that Mulligan. We bin getembad all the time there. No more this time now. This time nothing. I don't see im this time. *Janaka* I no more see im now, this time. Might be that bullock eat im, you know, finish im up. Because he likem eatem im, you know.[4]

Doug Campbell considered the idea that with the loss of prevalence there was also a loss of the knowledge of getting and cooking the food. *Janaka* is lightly singed in the flame of the fire and peeled open. The pith is the edible part:

---

3   Doug Campbell, notebook 38, 7–10.
4   Dora Jilpngarri, tape 82, recorded at Yarralin, 18 July 1986.

And *janaka* he's growing in the black soil. Oh, grow about that big, and early days people bin always get them. All them old girls bin always get them, and just get them out, and burn im on the light [fire], you know like that, just burn a big mob like that. Just work im up like that. And just getem out and just split im like that and break im like that and you get that part of it, inside one. Oh, bloody good tucker too!

Every way im bin grow here. But I can't see im growing now. Oh, im grow, but they don't get im now. I don't know why. *Janaka* he's grow any way long black soil, up here. Oh, after the wet he'll come down. You have a look around and I'll find one. He's still growing but they don't know how to use them now. Them olden time girls they know pretty well all that, only when we was born, oh, them young girls bin always chasing all that. Put im in the coolamon and bring im back longa that station, and cook im … and they break im like that, you know. Well you see that stuff coming out. Holy Christ, he's good one. Mmm. They bin always put im la coolamon, getem all that one everywhere, just use im, or you want to squash im up, you can eat im … like a damper or johnny cake. Sweet one too.[5]

Dora and Doug were born about 1912 and 1913. They remembered a time where there was so much *janaka* that people brought home coolamons full of it. Old Jimmy (born about 1905) also remembered a time when *janaka* was a staple:

I bin live longa that tucker *janaka* too. Limbunya Country, yes, when I bin little boy. I bin always cook im. *Janaka.* Im bin always … broke im, pullem, get that inside tucker, make im like a johnny cake.[6]

It took several years to find *janaka* (also called *janak, Abelmoschus ficulneus*). In a year of good rains we found it mixed amongst the sesbania pea (*Sesbania cannabina*), which is an increaser species on black soil pasture after very heavy stocking (Petheram and Kok 1983, 211). Once it had been a staple

---

5   Doug Campbell, tape 86, recorded at Yarralin, 24 July 1986.
6   Old Jimmy Manngayarri, tape 110, recorded at Yarralin, 13–14 August 1991.

for humans; it had been nearly eaten out by cattle, and now was surviving in amongst the weeds that take over after the cattle have eaten everything else. People born more recently remember *janaka,* but not as a staple. It seems to have been lost as a staple in the first few decades of the twentieth century.

One other plant that people spoke of with deep regret is called *kanjalu.* It once grew along the edges of springs and other sites of fresh water, and is a bulb, probably similar to a water chestnut. A site for *kanjalu* is in Old Jimmy's Country, at Kunja Rockhole on Kunja Creek:

> *Kanjulu* only longa Kunja rockhole … That's the only place. He all gone now. You know im might bin dry and finished.
>
> Debbie: We can't find that one any place.
>
> Jimmy: Yes, they want to find im that one *kanjalu* now. Good tucker. Good size, like this. And you cook im, cook im in the fire, oh, beautiful to eat. Oh, good.[7]

Figure 8.6. Kunja Rockhole on Limbunya Station, a Dreaming place for the water plant *kanjalu*, now believed extinct throughout the district.
Source: Photograph by Darrell Lewis.

7   Old Jimmy Manngayarri, tape 111, recorded at Daguragu, 14 August 1991.

Another plant that grew near sources of fresh water is *yarkalayin*. I have not been able to identify this plant either.[8] Daly explained that it used to grow back up in the rough Country around the springs and creeks of the watersheds. He thinks it may still be there, but it is no longer in any accessible area. He compared the loss of *yarkalayin* with the loss of *jarrwana*—both were formerly prevalent and now are difficult, or impossible, to find: '*Yarkalayin* back to Broadarrow Creek now. Some of them, I think. Oh, properly bullocky bin eat them. *Jarrwana* same. *Jarrwana* always been everywhere. Everywhere. Right up Timber Creek.'[9]

People spoke of former areas of abundance as 'farms' or 'gardens'. They were not proposing an analogy with cultivation, per se, but rather an analogy with knowledge and abundance. The garden is a site of abundance, and it is also a known site. Similarly for these hunter-gatherer peoples, the world was not randomly filled with food, but rather food was localised and predictable for those who knew where to go, when to go, what to look for, how to harvest it and how to prepare it for consumption or storage.

Figure 8.7. Severe erosion at a site that was once a 'garden'. In the foreground is a seed-grinding millstone. Gordon Creek, VRD, 1984.
Source: Photograph by Darrell Lewis.

---

8 Smith et al. (1993) have identified this plant as *Aponogeton vanbruggeni*—eds.
9 Daly Pulkara, tape 80, recorded at Lingara, 15 July 1986.

As Doug Campbell said, people would leave some there for next time; food getting was not a matter of harvesting it all, but rather of taking a portion and leaving a portion. Jessie Wirrpa contrasted this human behaviour with the behaviour of cattle. In commenting on the huge reduction in the amount of 'bush bananas' (*Leichhardtia australis*), she said that cattle eat the whole vine, and eat the fruit before the seeds ripen and fall, so the plant has no way to reproduce itself.

We were able to document many more plants than people actively use these days, and one of the reasons for the lack of use is that plants that are still living in the area are greatly reduced in numbers and are growing sparsely. They live in out of the way refuge areas rather than in their own proper habitats. This is especially true of plants whose preferred habitat is where cattle graze, and of course, of plants that are grazed by cattle, horses, donkeys and other introduced herbivores. The best sites for collecting specimens were around the bases and up into the crevices of sandstone ridges. In these areas the cattle rarely ventured, and many plants whose proper habitat is not a sandstone crevice were nevertheless hanging on to life in these places. In terms of the continuing presence of living plants, the existence of refuge areas preserves diversity for the future, but in terms of making effective use of the many foods, medicines, tobaccos and technological items which are now out of place ecologically and reduced to the point of extreme scarcity, the situation is not workable.

Daly Pulkara thought about where things were growing, and how they were surviving. Shortly after our conversation about 'the wild' he had the opportunity to fly into Country he had not visited for decades. He developed a set of contrasts which constituted a continuum ranging between quiet Country on one hand, and various stages of degrading Country on the other hand:

> I went up there with a plane and find that Country [inaccessible 'quiet' Country which was being documented under the Aboriginal Areas Protection Legislation]. Oh really lovely, good place. Just like the time when we bin there [footwalk]. My Country look nice and good yet. You know, my father walking there before, well he [Country] was looking same yet. Not scrubby or more waste. Here [at Lingara] we got a little bit, you know, we not look ugly here, but ... You go back to Yarralin, you'll see all kind of thing. Rubber bush [an introduced weed] never grow

up to this Country yet. That coming from different place. [At Yarralin] More scrub. More waste. You looking back to Lingara, here, you look everything true every way. Same tree, anyway. Yeah, yeah, I bin properly glad to see that.[10]

His qualification 'same tree, anyway' in reference to Lingara is his acknowledgement of a specific loss at Lingara. The trees are there, but the grass is gone. Lingara is an outstation that was established on Humbert River Station after the pastoral strikes. The Humbert River mob (Daly, Snowy, Riley, Nina Humbert and others) had walked off the job in 1972, and when they returned, they came to Yarralin on VRD Station (Rose 1991, 233). They had never stopped thinking about a community of their own in their own traditional Country on Humbert River, and in 1980 they established an outstation at the site of a Dreaming tree that is an increase site for the seed-bearing grass known as *ngaruyu, mangorlu* or *lingara* (probably *Fimbristylis oxystachya*). A Dreaming tree identifies the area of the Grass Seed Dreaming.

The seed-bearing plants are so rare that they are almost gone. Grass seeds were, of course, both a staple and a strong identifier of women's work, mother's care and Country's bounty. It was still an obtainable food in the 1940s, and Doug spoke of bags of seeds being sent back to VRD. I asked Kitty Lariyari about *mangorlu* (*ngaruyu*): '*Mangorlu* too. Oh, this time nothing now, this one *lingara*. This time nothing now. No more like before. Oh, too much [plenty] im bin there.'[11]

The site that should promote abundance no longer does so, and the meaning of the place has become a memory for the older people, and a story out of the past to younger people. The time of grass seeds once was a time that unfolded into the world from season to season following the rains. There is a little seed-eating bird whose song tells the rain to go away and heralds the arrival of dry weather and ripening seeds. Time, in this context, has stopped flowing, and the ripening seed event that is linked with the bird's song to send away the rain no longer occurs. Loss of a significant species (whether that loss is an outright extinction or the loss of adequate abundance) is a loss of time, continuity and communication.

10   Daly Pulkara, tape 80, recorded at Lingara, 15 July 1986.
11   Kitty Lariyari, tape 85, recorded at Yarralin 24 July 1986.

Faunal loss

Reptiles seem not to have been so badly affected by the intrusions of European domesticated animals, although skinks and other reptiles are preyed upon by feral cats (Low et al. 1988, 16). Doug Campbell remarked upon the decreasing numbers of goannas when I asked him if he had never been hungry in his life: 'Never for nothing in my life. But this time you can't see em anything goanna walking around. You can't see em anything grow, nothing. You can't. Can't find nothing.'[12]

With mammals the story of loss is extreme. As stated, scientists estimate that up to 60 per cent of the animal species in the arid and semi-arid cattle country may be locally extinct. I cannot speak with that kind of certainty for this area, but there is no reason to suppose that it is any better than elsewhere. There are four main losses that my teachers discussed with me: bandicoot (*puluka; Isodon auratus*), bilby (*jarkulaji; Macrotis lagotis*), brush-tail possum (*jangana; Trichosurus arnhemensis*) and 'native cat' (*parjita; quoll, Dasyurus hallucatus*). Several of the oldest people had seen all of them in their lives, but no one had seen any of these animals for decades. A study commissioned by the Conservation Commission of the Northern Territory claimed to have identified signs of bilby in the Bilinara sandstone (Low et al. 1988, 14). It thus seems possible that the animal still survives in the region, but Aboriginal people have not seen it for years.

I found it impossible to make strong identifications of mice, rats and other small marsupials. It seems that a number of them may have been classed together under one term, but in the absence of specific individuals, and in the absence of recent and detailed knowledge, it was not possible to determine which animals were still in the region and which were no longer there. The small delicate mouse (*Pseudomys delicatulus*) and the pebble mound mouse (*Pseudomys johnsoni*) are still to be found. Others undoubtedly were there, and some may still be there, but I have not been able to make identifications. The Low ecological survey found evidence for nine native mammals on VRD Station: agile wallaby, northern nail-tail wallaby, greater bilby, red kangaroo, common wallaroo, antilopine kangaroo, little red flying fox, hoary bat and the common planigale (a carnivorous marsupial mouse).

---

12   Doug Campbell, tape 87, recorded at Yarralin, 25 July 1986.

Many of these absent animals are the totemic relations of people who are still alive. The people live, and their totemic sites, songs, designs and practices live, but the animals themselves, the non-human descendants of the ancestral totemic figure, are gone. This is the case with Jessie and Nina, and their possum totem in their mother's father's Country. The site is there, and it contains a substance that brings good luck to women, so in that sense the possums are still active in the world. But these relationships too are diminished and diminishing. Jessie and Nina, with others, got the protection of the Aboriginal Areas Protection Authority for the possum site. They said that if the area were to be bulldozed or otherwise damaged so that the site and the mineral deposit were damaged, even more power would be lost to them and the Country.

Similarly, one of the major sites for Bilinara men is a hill (Mount Northcote) that is sacred to the northern quoll, or native cat. The senior man for this site, Hector Wartpiyarri, had seen this animal when he was young. He knew what he was talking about, and he also knew that it is not there anymore at all. Old Tim described the *parjita* as 'some sort of pussy cat, but he's devil now'. Old Tim used the term 'devil' as a gloss for what other people often termed 'spirit' (in the sense of 'ghost'). Devil or spirit is the living presence of something that is otherwise gone from the world of the living. Old Tim's words tell us that all these sites, ceremonies and relationships are with a creature who now only lives in the world as spirit, ghost or devil.

Big Mick Kangkinang (born about 1900) remembered the animal called *wirimirimawu* in Ngaliwurru language. It is a gliding possum, and its southern range once coincided with the southern side of the Stokes Range, in the same area where the boab trees and the *kakawuli* (yams) reach their limit. When Darrell Lewis and I were working on the land claim for this area we needed to know what a *wirimirimawu* was because there was a Dreaming site there. Big Mick described the creature, and the identification was clear. He was the only one who knew. When we told other people what Big Mick had said, they suggested that he was getting senile and did not know what he was talking about because there is no such thing. When we said that such animals really do exist further north, they agreed grudgingly that that might be possible, and they accepted Big Mick's word for the identification of this creature. Here again, people said that whatever it was, it was only spirit now. There was nothing more for that Country, and hence, in their view, nothing more at all of *wirimirimawu* in the world for them.

Finally, one can also consider some of the connections here. In Kayalarin Country, in the big black soil plains just south of the Stokes Range, there is a Dreaming site for possum and native cat. They were there together pounding *kayalarin*. Now the animals seem to be gone, and the plant is gone. The site is there, but what does it reference?

Billabongs and springs

A long-term white resident of the Victoria River District who has observed changes in the land and the waters over many decades states that erosion is severe and seems to be unstoppable. For example, earlier efforts to slow down erosion by throwing old tyres and old cars into erosion gullies are now known actually to speed up the process. The rivers are all becoming wider and shallower, and homesteads or communities built on the banks of rivers could go under one day. This perceptive person had observed that many natural waterholes have dried up. Indeed, the pastoralists' practice of bulldozing or dynamiting springs to try to get them flowing again has probably reached the end of its feasibility. The springs are drying up, and the waterholes in the big rivers are also drying up. It seems that the aquifers are not being replenished as rapidly as they are being emptied.

The results today hark back to the earliest days of settlement when one of the most contentious issues between settlers and Aborigines was water. Settlers wanted it for their cattle, Aborigines needed it for themselves. Most native animals also required water. There was an undeclared resource war fought over water, with tragic consequences for all, including water.

One of the few white men killed by blacks in the Victoria River Country was Jim Crisp.[13] The event was relatively recent (1919); it took place on Bullita Station, just north of Lingara. The killing was in retaliation for Crisp's murders of Aborigines. Riley Young explained that Crisp had been shooting Aborigines because he thought they were 'buggering up the water'. He saw them using fish poison, and he interpreted the incident as 'buggering up'. He may not have understood that the poison would have no long-term effects, or perhaps he took the view, as so many did in those days, that the Country (perhaps even especially the water) was there first and foremost for cattle and that the Aborigines would have to go. In any case, Riley says

---

13   Lewis (2012) documents the deaths of 15 settlers killed by Aborigines in the Victoria River District. A much greater number of Aboriginal people were killed by white men—eds.

he shot people over water, and they killed him in return. At least one and possibly two men identified as the killers were later shot by police (Schultz and Lewis 1995, 54) but any massacres following the killing are unrecorded in European documents (see Rose 1991). Riley also liked turning people's words around on them, and his response today to Crisp's reported allegation that Aborigines 'buggered up the water' was that they were walking for years and years and always 'buggered up' the water, because it was their water.

The discussion of the Rainbow Snake (Chapter 5) led Riley to talk about other parts of his Country where the surface and underground waters are receding, and he linked the loss of water to the loss of people. For me, few things are more indicative of 'buggered up' water than the loss of lilies. Racecourse Billabong close to VRD Centre Camp was a resource site for Dora and Doug when they were little, and it was full of lilies. Numerous other billabongs and springs held lilies, and every once in a while some of them still do. Others, though, are eaten out, and some have been empty of lilies for decades.

In earlier years the Pigeon Hole mob used to send bags of lily corms to VRD people. The lilies in the billabongs there came from Dreamings—some from a goanna who brought them from the river, and others from the Nanganarri Women. In addition, there is a Lily Dreaming at one of the billabongs.

Figure 8.8 shows a Dreaming site for lilies near Pigeon Hole. The stone is the source for water lilies (*mintarayin, Nymphaea macrosperma* according to Wightman 1994, 40). The lilies at this billabong are believed to have been placed here by the Nanganarri Dreaming Women, and the stone contains the life and Law of lilies. Anzac Munnganyi was striking the stone with green leaves; this is his Country, and it is his work to perform this ritual. In this case, however, his action was simply a demonstration the purpose of which is the proofing of evidence for a claim to land under the *Aboriginal Land Rights (NT) Act 1976.* Anzac explained the meaning of his actions: that this is a lily site, that the 'proper really tucker' comes from here. You sweep the stone with green leaves, and that pushes the lilies back to the billabong. You talk to the Dreaming, saying 'You make plenty tucker'. After the wet a big mob of tucker will come up. The (Dreaming) stone puts the lilies there in the billabong. Anzac was asked if he had a song for this place and he said, 'No song, just the words. Really true words.'

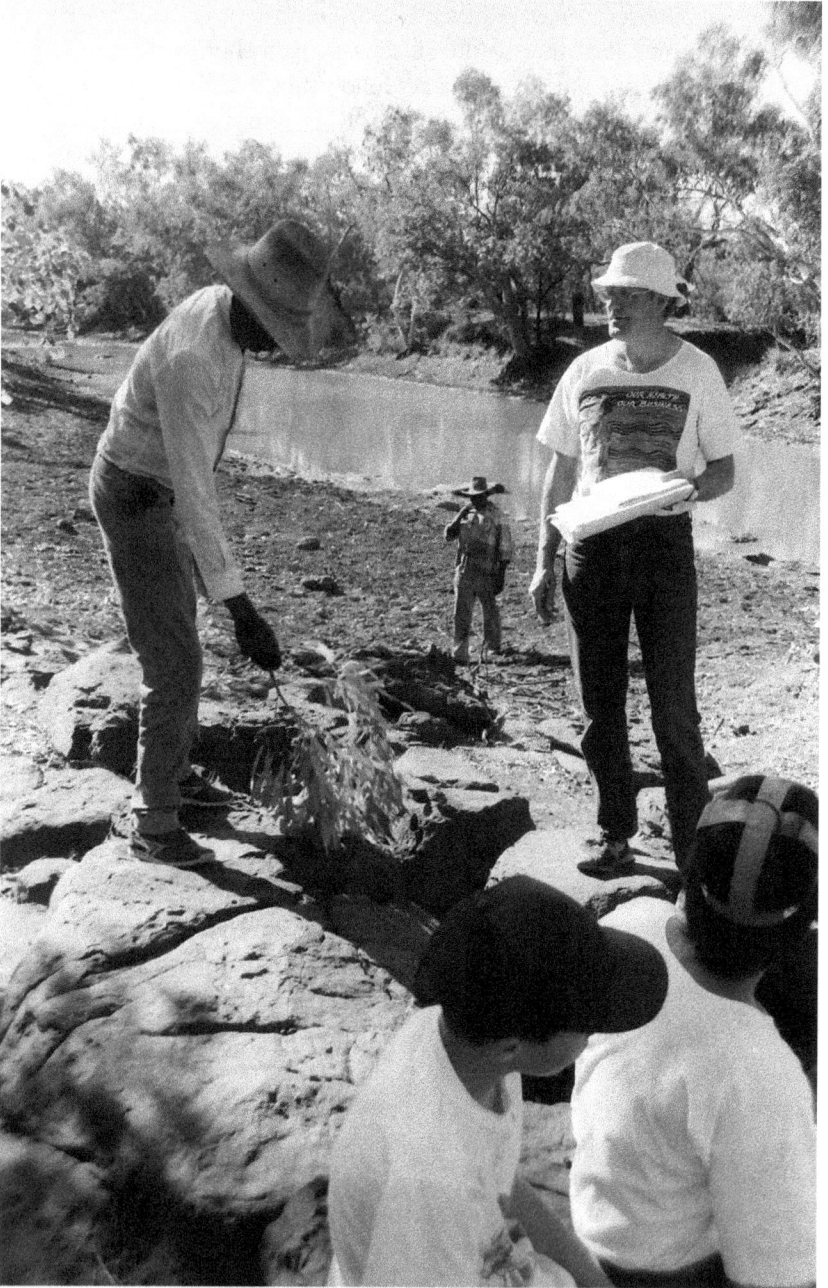

Figure 8.8. Anzac Munnganyi performing an increase ritual by brushing a Lily Dreaming during the Pigeon Hole land claim hearing, 1988.

Source: Photograph by Darrell Lewis.

If you look closely at the background of Figure 8.8 you will see a muddy billabong surrounded by trampled mud, but you will not see any lilies. This Country has been grazed by cattle for over one hundred years. The lilies disappeared in the 1930s, as near as I can determine. There are two closely related billabongs, and Hobbles explained that the traditional owners had been able to bring the lilies back to the other billabong, but not to this one. Rituals for lilies for this billabong are no longer performed, as it is believed to be a hopeless case under current land use patterns.[14]

Floral invasions

There is a huge number of introduced plants. Some, like the introduced pastures, have come in as replacements for Indigenous plants. Others are escapees from homestead gardens. Some rode in on the backs of camels, and some have arrived in trucks along with loads of hay (see Lewis 2002, 41–42). Those that survive are opportunistic: either they take advantage of disturbance in order to colonise ground where natives are unable to take hold, or they find conditions to be so favourable that they start to drive out natives. Under the Northern Territory's *Noxious Weeds Act 1963,* weeds are classed into one of three categories. Class A weeds are to be eradicated: they pose a significant threat but occupy a small area so there is good chance of eradication. Class B weeds are to be controlled: they are widespread and thus it is impractical to eradicate, but prevention of further spread should be possible. Class C weeds are not to be introduced to the Territory. All Class A and B weeds have the status of 'declared noxious weeds' in the Territory (Miller and Crothers 1998).

Although pastoralists are obliged by law to eradicate or control these plants, the efforts of those who make the attempt (and not all do) are unsuccessful. As is well known, eradication or control of noxious weeds must be done at a regional level. If there is not the political will to force pastoralists to take effective action, and if there is not the will to dedicate the kind of money that would be required if the government were to undertake the task, many weeds will remain uncontrolled.

---

14   In the manuscript, this section was followed by the heading 'Desertification', but there was no text—eds.

Daly spoke of 'rubber bush' as one of the plants that makes Country 'ugly'. *Calotropic procera* grows along all the roads in the district, as it thrives on disturbed soil. It is a declared noxious weed. Others abound. Perhaps the most disturbing at this time is that the weeds are now affecting the banks and beds of the Victoria, Wickham and other rivers of the region. Castor-oil plant (*Ricinus communis*), Noogoora burr (*Xanthium chinense*) and devil's claw (*Proboscidea louisiana*) are all declared noxious weeds; castor-oil plant and Noogoora burr have taken over long stretches of the banks of the big rivers.

The immediate effect on Aboriginal people is to make fishing very difficult because of the difficulty of river access when one has to walk through painfully prickly and clingy weeds and sit amongst plants, especially *Ricinus communis,* whose berries contain one of the world's most powerful poisons (ricin). Along with fostering that sense of disheartened anguish that I mentioned earlier in connection with lost species, the actual difficulties of getting to the river mean that people are spending less time fishing. The diet is impoverished and so are relationships. Yet another bush tucker of the kind that Country 'gives' its people is slipping out of the repertoire. I am advised by scientists that these and other noxious weeds are driving out native vegetation. This means that the foods that support the fish and turtles are diminishing. In due course (and perhaps extremely quickly), there will be a collapse in the fish and turtle populations of the larger life-sustaining rivers.

It would be nice to think that everyone would have an interest in keeping rivers healthy and accessible. However, with pumps, bores and the seemingly unlimited access to aquifers, it may be that pastoralists have little interest in ensuring the ecological stability of the rivers. Indeed, there may be a feedback loop here. In the 1980s pastoralists were urged to fence off river access in order to keep the cattle away from the rivers so that the degradation of the banks, and the accompanying siltation, could be curbed. Large stretches of the Victoria River were fenced off. Did cattle help to keep down the noxious weeds by trampling or even eating young shoots? The evidence is not all in, but this suggestion has been made further north where the invasive *Mimosa pigra* erupted into plague proportions when the buffalo were shot out of the country. Here, too, scientists have been reluctant to draw firm cause and effect relationships, but the correlation is suggestive (Walden et al. 2004, 12–13).

One type of floral invasion is the spread of 'woody weeds' that almost certainly is due to the cessation of Aboriginal burning and changes in stocking regimes. Common woody weeds in the Victoria River District are acacia species which can be controlled with proper burning, so the issue is not how to control, but whether there is a will to control. That will is unlikely to emerge until acacia scrub significantly impacts on profitability. For my teachers, encroaching scrub is blocking up the Country.[15]

Riley contrasted the state of the Country before the 1970s with its current state:

> And when I been go long this land, land was really good. He was really good ... But now ... when we been start again, Country was little bit funny that day, I been looking at Country was little bit funny that day. I been looking at, 'What's wrong this one? Something wrong.' And after that I been look now, one year's time I been see em plants been get up. You go longa bush now looking for fruit, you can't see em fruit. You see em all these trees now ... And even if you go round la bush here, you can't see that *karil*, gooseberry, *kilipi, tipil, purlkal, ngaringari,* that kind been too much longa this—*yarkalayin, mintarayin,* that been already been clear. But you can't see em this time now. That's from what I been say: 'Country been change. Ground been change.' Because no fruit now. Where him gottem good fruit longa this ground, Country was look good. But fruit going away, Country gone, finish.[16]

Indigenous Country is not just 'nature' in some neutral sense. It is Country, it is creation, it is kinship, history, relationship and the future. Invasive species, therefore, invade kinship, history, the future, creation and much more. They generate huge ramifying issues that Indigenous people all across

---

15   A note here indicates that Debbie intended to add a summary from Lewis (2002)—eds.
16   The plants Riley discusses are: *karil: Cucumis* spp., probably *trigonus;* gooseberry: *Physalis minima; kilipi: Leichhardtia australis; tipil: Antidesma parvifolium* (unable to locate a specimen); *purlkal: Vitex acuminata; ngaringari: Pterocaulon serrulatum; yarkalayin: Aponogeton vanbruggeni* (unable to locate a specimen); *mintariyin: Nymphaea violacea.* Debbie intended to include here discussion on global climate change—eds.

Australia are really struggling with. Aboriginal people have grasped the existential issues of massive landscape alteration far better than mainstream Australians have.[17]

An 'NRM' detour

In the public arena where 'nature' is debated, the prominent discourse of conservation and care is a discourse of management. In Australia, management is fastened down by the acronym NRM—natural resource management. NRM is explicitly or implicitly goal-oriented, and these days the goal is some form of sustainability. Types of sustainability clash, so there is a meta-goal of sustainably balancing conflicting types of sustainability—economic sustainability to be sustained without impacting on environmental sustainability, and both to be achieved within a society that has a stated goal of moving toward increasingly sustainable forms of social and environmental security. My personal favourite is the sustainability goal articulated by the Cooperative Research Centre for the Sustainable Development of Tropical Savannas (CRC-TS). In contrast to the name, the stated goal is sustainable habitation (of tropical savannas), and in this northern frontier context the goal seems to express an optimism the very necessity for which is scary: they hope that Australian people will be able to continue to live there.

There are many routes into a critique of NRM; mine starts with the unsustainable separation of nature and culture. This implicit platform for management is ecologically mistaken and is socially inept in that it excludes from the equation any mention of the very species that purports to be doing the management. The concept of 'resource' excises elements of the 'natural' world from their context and highlights them as if their main reason for being was to provide services. It implicitly or explicitly denies that 'nature' has its own ontological status. The concept of 'management' implicates the human, but mistakenly implies that it is possible for a subsection of one species to gain sufficient understanding of the context within which it is embedded to enable it to make good large-scale decisions about its own context.

---

17   Debbie had the following note here: 'Add restoration (ecological restoration) and discuss re-introductions in central Australia; requires different type of land tenure, different concept of production, different value of labour—that taking care of Country is productive labour. It probably hasn't much hope if in competition with pastoralism?'—eds.

Analytically, NRM looks like a disaster. The larger problem lies in the disjunction between short and long term. NRM is absolutely essential in the short term. Every day, every hour, people are making decisions that have long-term environmental consequences. There must be some guiding principles for how such decisions are to be made, and some goals toward which such decisions aim. In the short term, some forms of sustainability constitute reasonable and desirable goals. Increasingly, local knowledge, social goals and conservation management that cut across different types of land tenure all stand to undermine the strong dualisms signalled by the master acronym. NRM is in some contexts subverting its own epistemology and is definitely deferring the 'reckoning' (Athanasiou 1997).

In the long term, NRM works toward disaster because it is embedded in a set of epistemological errors, the consequences of which are ramifying and recursing exponentially. Short-term necessity and long-term disaster: NRM might be described as a self-fuelling arena of impending implosion.[18]

---

18   Debbie had a note here saying, 'Add the stuff about non-management—being called into action, Country telling you where to go and what to do (from forest lecture).' She also said, 'Add something about our species—we have to take seriously our huge capacity for damage; at the same time we need to take seriously the resilience of living systems, and the positive benefits that might derive from leaving things alone.' The 'forest lecture' referred to is Rose (2002)—eds.

9

Coming into Life

We'll run out of history, because *kartiya* fuck the Law up and [they're] knocking all the power out of this Country.

Daly Pulkara

Death and life

Stephen Muecke ventures a generalisation with which I agree: Aboriginal philosophies 'are all about keeping things alive *in their place*' (1999, 34). One of the motivations that impelled my teachers to take enormous amounts of time in remembering the past, naming plants and animals, discussing the details of the lives of plants and animals, and teaching a myriad of other aspects of life in Country was that they wanted to document all that has been or is being lost. They wanted conquest to be known to have effects that have rarely been acknowledged, much less addressed, by white fellows. Their desire was shaped by the ethic of self and other that I have discussed previously. That is, Aborigines thought that if white people understood how knowledgeable they were, there would be greater recognition of land rights, of rights to be involved in land management decisions, and other matters such as basic respect. At the same time, they held the view that memory was required because so many of the living things to be remembered no longer seemed to exist. That is, if the living things were there, one would not need memory, as one would have experiences of encounter. Presence would be face-to-face; memory picks up where presence leaves off.

In assessing white fellows, many of my teachers over the years made statements that led me to develop the phrase 'epistemological chaos': white people, my teachers seemed to be saying, were in a state of chaos, not knowing what to remember and what to forget. Further, they followed dead laws and failed to honour living laws. And further, in their power and arrogance their failures led them to promote death and to fail to honour life. Thus, memory and forgetting were intimately connected with life and death, and with the newly troubling issue of absence that is, more than elsewhere, presence. These are the connections I wish to explore in the context of time concepts.

I have discussed aspects of these issues elsewhere (1992, 203ff, 2002) and I will review the main points briefly. Victoria River Aborigines frame a period of about one hundred years as the time of ordinary human remembrance. This is a period within which the genealogies are set, within which people tell stories about things that happened to their parents and perhaps grandparents, and within which people's own life histories are set. Along the past horizon of this period, memory works by packing the present away. The main forms of remembrance are stories and songs, associated with designs and performances, that are all focused on place. Memory is thus kept available to the living by being worked into the geography of place and becoming part of the stories and songs of that place. Stories and songs do not stretch out infinitely to accommodate the lives of every generation that ever lived. To the contrary, in this system of remembrance much will be forgotten. Only events that prove to be significant over the duration of this long period will be subjected to memory work.

It does not follow, however, that that which is forgotten is thereby obliterated. To explain how the past can be packed away into non-memory without being obliterated, I consider the intersection of two time concepts and the place-based origins on which they depend. The first time concept is the ecological time in which life comes into the world. Two of the main characteristics of ecological time are that it is sequential and it is irreversible. Life comes into the world in sequential fashion (children follow parents, for example), and the dynamics of the world are such that in a seriously alive system, this is an ongoing process. The second point is that sequential time is irreversible. Living things grow and die. That is their fate.

The second time concept is that of boundedness. Individual life emerges into sequential time, lives and dies. Ecological time continues to unfold (I will discuss its fragility later). Individual time is bounded. Both are irreversible.

Ecological time coexists with and depends on creation or Dreaming. There would be no unfolding process of new life if there were not a process of life's coming into the world. The paradigm of coming into life centres on birth, as I will discuss shortly. For the moment we will confine the discussion to creation. Creation is linked to place and is ongoing. It is not sequential, irreversible or unbounded, but rather is the continuous happening of emerging life. The linking concept between ecological time and creation is the return. A good example of the intersection of ecological time and creation is the life and death of a person. As discussed in Chapter 1, when a person dies their 'spirits' go in several directions. One spirit returns to Country and joins the dead, who take care of Country and take care of the living. Another spirit joins the flow of ecological time and is reborn either directly as a new person, or indirectly through animals to human personhood. Sometimes the return is named and known, other times it is not; but even if it is not known, it is understood to have occurred, as this is how life works. The dead immerse themselves in the flow of life, and keep returning to life, often in their own Country. The dead are forgotten, in the long term, but they are not obliterated: they keep returning into life.

This culture of life in place contrasts with the western culture of death, as exemplified in Christian iconography and belief, and as analysed in secular and philosophical terms by the eco-philosopher Val Plumwood (1993, 97–103). She offers an excellent account of the west's relationship with death and concludes that from Plato through modernity our culture has failed to offer a life-affirming account of death within the domain of this world. Either it has sought to resist death, or it has sought to redeem life in an afterworld that is dissociated from this world. In contrast, Victoria River Aboriginal culture claims that death does not have the last word in this world. Death is of course grievous both to the one who experiences it and to those who have to live with the loss. Death does constitute a loss of the particular person that is irreversible. The 'spirit' keeps returning, however, and the dead do live on in their own Country. In both of these contexts, loss is not absolute, and life has the last word.

Colonising violence has a devastating effect, not only because of the massive and irredeemable death that it caused so wantonly and indiscriminately, and not only because death, as I have shown, is everywhere, but also because of the last word. In the extreme cases of extermination, death has the last word to the extent that what is gone is probably gone forever. Not only does death have the last word, but in many instances the conquerors, too, claim the last word. A modest example is the scientific inventory of currently living species in a particular place. The inventory becomes complicit with death when it purports to be comprehensive. If taken as comprehensive it erases the lives of plants and animals who once lived there but are now gone forever from the place, or even extinct.

Often the last word in conquest is denial of much of the death that has been caused. Riley Young told a story which I have written about in detail elsewhere (Rose 1991, 259–60). He tells of an Aboriginal man who was required by the station white fellows to collect a huge pile of firewood. The man did not know, but as Riley tells the story he makes sure that we know, that the firewood was going to be the pyre of the man who was required to collect it. After he had collected a big enough heap, the white men asked him to stand before the pile for a photo. They asked him to smile, and as one man clicked the camera, another fired a gun. The man was killed and burned; what remained was a pile of ashes and photo of a smiling black man. One of the many appalling things about this terrible story is that the cunning and cruelty of white fellows encompasses, disfigures and disguises death. The fire disfigures death, leaving no body for the survivors to mourn. The photo disguises death, representing it falsely as life. If Riley's story were lost, very little would remain; most prominent, perhaps, at least in the public record, would be the photo of a happy blackfella smiling as he takes off his hat, not knowing that he is about to die. Only memory prevents the camera, and the conqueror, from having the last word.

Riley's story was not the only one founded in relations between the living and the dead. Another of my teachers, Old Johnson Bididu, told me that he stayed at Lingara, which at that time was a small community with marginal land and uncertain tenure, because there they camped on the blood and bones of their people who had been murdered. The evidence that white people sought to disperse through fires or drownings, or unmarked graves, or through silence and lies, has for Old Johnson and others become part of the place. Memory, place, dead bodies and genealogies hold the stories that tell the histories that are not erased, and that refuse erasure. Painful as they are, they also constitute relationships of belonging, binding people into the Country and the generations and the knowledge of their lives.

I am pressing us to think about how these stories configure life and death. Western culture pervasively imagines the relationship between life and death as a battle. In this battle the grave will always claim at least a temporary victory, as death is inevitable, but numerous cultural practices seek to redeem life. The conceptualisation of death as sacrifice is foundational not only to theology, but also to society and nation (Muecke 1999). Paul's triumphal assertion to the Corinthians that through the resurrection 'Death is swallowed up in victory' sets out the matrix of the war with death (I Corinthians, 15:54, RSV). In the contemporary world, historians are deeply implicated here, as Anne Curthoys and John Docker (1999) note in their analysis of nineteenth-century history's desire 'to defeat time and death'. History, they contend, is a continuing act of defiance.

My Aboriginal teachers know the pain and grief that death entails, of course, but the grave is not given victory. In part this is because life is not at war with death, but more significantly, a culture of life promotes relationships, connectivities and patterns that sustain life. Living things emerge, and continue to emerge, for as long as life has the last word. As often as they are erased or exterminated, the sting of death walks the land. As long as death amplifies itself so that living systems are impeded from being self-sustaining and self-repairing, death carries on the work of obliteration.

Creation

As previously observed, the Australian continent is crisscrossed with the tracks of the Dreamings: walking, slithering, crawling, flying, chasing, hunting, weeping, dying, giving birth. Performing rituals, distributing the plants, making the landforms and water, establishing things in their own places, making the relationships between one place and another. Leaving parts or essences of themselves, looking back in sorrow; and still travelling, changing languages, changing songs, changing skin.[1] They were changing shape from animal to human and back to animal and human again, becoming ancestral to particular animals and particular humans. Through their creative actions they demarcated a whole world of difference and a whole world of relationships which crosscut difference.

Where they travelled, where they stopped, where they lived the events of their lives, all these places are sources and sites of Law. These tracks and sites, and the Dreamings associated with them, make up the sacred geography of Australia; they are visible in paintings and engravings; they are sung in the songs, depicted in body painting and sacred objects; they form the basis of a major dimension of the land tenure system for most Aboriginal people.

---

1   Minimally this term 'skin' refers to social categories, the English technical terms for which are section, subsection, semi-moiety and the like (depending on the precise organisation of the skins).

To know the Country is to know the story of how it came into being, and that story also carries the knowledge of how the human owners of that Country came into being. Except in cases of succession, the relationship between the people and their Country is understood to have existed from time immemorial—to be part of the land itself.

The relationship between other species and Country is that they too belong there because they have their origins in Dreaming. Other species act as they do, communicate as they do, live where they do, and interact as they do because Dreaming made them that way.

The life of the flesh is the life of the spirit, and the Law that brings life into being is a Law that is managed by women and by men. The major rituals, said once to have been held exclusively by women, are now shared between men and women. As well, there are, of course, many portions of restricted ritual which are participated in by both women and men; there are rituals which are managed by women and are also carried out in the presence of men; and there are rituals, or portions of rituals, which are exclusive.[2] The organisation of ritual parallels the organisation of geography: there are places which are managed jointly; places to which men may go but are not allowed to know the meaning of; and places where men can never go. In this physical landscape, all of which is spiritual, there is women's space and there is men's space—absolutely.

Male Dreamings, too, imprint themselves on the Earth, and leave behind the traces of their activities, the sites of their actions, and their specific presence. When the salt water pulled back, as Victoria River people say, life emerged, or was 'born' from the Earth. Some of this life was male, and some was female. Males and females, whether flying foxes, or possums, or human beings, travelled the Earth creating a gendered landscape. The land does not privilege women to the exclusion of men, nor does it set women

---

2   Women's ritual life is described and analysed in detail in Diane Bell (1983), Catherine Berndt (1950), and P.M. Kaberry (1939). There has been some discussion about the content of women's ritual. Bell emphasises Country and nurturance as key foci; Berndt emphasises sexuality and the control of men. Annette Hamilton (1986) criticises Bell's emphasis (see also Hamilton 1981), as does Francesca Merlan (1988a, 1988b). In my work I have seen less nurturance, but intense focus on Country, Dreamings, sexuality and the control of men. It seems to me that women's ritual ('business') is broad enough to accommodate the range of women's life issues in their historical contingency. No single focus ought, I believe, to be taken as the defining feature of women's ritual life. Peggy Brock's (1989) edited publication presents a number of essays which indicate a range of approaches and issues. Elizabeth Grosz (1989, 1994) provides an excellent analysis of the issues I explore here; as far as I know, she was the first to take Aboriginal women's religious practices truly seriously.

in opposition to men, although it does acknowledge the competitive quality of desire. Gendered land locates women spatially and cosmologically. The process of bringing life forth into the world is Indigenous, and it is female.

Daly Pulkara, one of the men who taught me, explained that this Earth, referring to the Country for which he is an owner within Aboriginal Law, has an Aboriginal culture inside. Other people made similar statements; for example, that everything—language, Law, kangaroos, trees—all come from the Earth.

Gender

Gertrude Stotz has offered what I believe to be one of the most powerful interventions in the study of gender amongst Australian Aboriginal people. Her research was carried out with Warlpiri and other desert people to the south of the Victoria River region. The exposition that follows is based on Stotz (1993) and is interpolated with my own evidence and further analysis. I can lay out the analysis first as a series of statements. Sex, as the distinction between male and female, is one boundary in defining and separating human beings. Country (mine, yours, others') is another boundary in defining and separating human beings. Countries are separated first by Dreaming creation, as discussed above. Each Country is associated with its own group of people, and each Country is exogamous. An individual has relationships of connection with the Countries of their ancestors: mother's (father's) Country, father's (father's) Country, granny's (mother's mother's) Country. Each Country has gender-restricted business that is for the Country, and for the people of the Country. The business for each Country requires descendants of women and descendants of men.

Stotz goes on to argue that within the gender-restricted domain of business a further gender division is effected. Women do the work of fathers in their father's Country and do the work of mothers in their mother's Country. Men, according to this argument (which must be partially hypothetical because neither Stotz nor I have access to that domain) do the work of fathers in their father's Country and do the work of mothers in their mother's Country. The argument is, thus, that each person is both father and mother in respect to the bringing forth of the life of their particular Country. Gender-restricted domains depend on sex, age and status in initiation. They consolidate domains of women, and domains of men, and the difference between the two is a difference that really matters.

As many scholars have noted, failure to respect the difference, to breach the boundary of gender-restricted business, is to risk life itself. There is nothing ambiguous about it at all. However, within that restricted domain, gender is re-ambiguated. Women are not only women, and men are not only men. Each is a generative and nurturant partner in the rituals that bring forth the fertility of Country.

Stotz contends that the procreative unit is 'mother' and 'father'. The ambiguation of this seemingly predictable fact lies in the nexus of Country. Each person acts toward their own mother's and father's Countries. Husband and wife have different Countries. In respect of every given Country, therefore, the procreative female and male are not 'wife' and 'husband', but rather 'sister' and 'brother' (because they share the same mother and father Countries). In fact, of course, gender-restricted business ensures that brother and sister will not act together as a procreative couple; each acts as mother or father within their own gendered domain. Their similarity is that they are 'mother' and 'father' for, or in, the same Country. In *Dingo Makes Us Human* (1992) I discussed some of the metaphors of birth and continuity in relation to bodily substances; my argument there is identical to the argument here. If relationships of connection are inherited from mother and father, and passed on to one's own children, but do not mingle and merge, then the separation of spheres of ritual action, and the ambiguity surrounding the parental contributions of substance to the new person, preserve the integrity of the difference between sister–brother and wife–husband.

Procreation

In the world of change, where everything is in the process of becoming different to what it was, one of the great philosophical questions is: What is the quality of change? Philosopher Lev Shestov (1970) addresses this issue in a remarkable essay entitled 'Children and Stepchildren of Time: Spinoza in History'. He contends that mainstream western philosophy, for more than two thousand years, has identified as one of its two main projects: to love that which is immutable and eternal. A love for the immutable is, in western philosophy, a lack of love for, or indeed a denigration of, the ephemeral world that lives through change. Thus, time and change are the poor relations of western philosophy, and so, too, is the world. As Shestov puts it:

> Those who have meditated on this question have established … so strict a bond between the idea of death and the idea of change that the two ideas at present are only one. That which changes now appears as insignificant, as miserable, as that which is condemned to die. (Shestov 1968, 406–7)

Shestov's argument is against this focus on death, and toward a focus on birth as the critical moment of our lives. He equates eternity with death and values birth as the entry of living things into time and motion (Martin 1970, 37).

As is well known, for decades western scholars debated Aboriginal people's understandings of conception. Did it take two to make a baby, and if so, which two? A mother and a father? A mother and a baby spirit? A mother and a Dreaming? Or did it take three? Mother, father, spirit? Perhaps four: mother, father, spirit and Country? Is it more complex than that? Most recently, scholars have started debating the history of previous generations of debates (Wolfe 2000). Given that much of the information that might resolve these debates is locked away in secret business, the perverse tenacity of the scholarship suggests that it may depend far more on western concerns about human knowledge than about Aboriginal people's declared and undeclared knowledge.

I do not propose to intervene with answers, as some of the information is not available for public discussion. I hope my questions have been sufficiently suggestive (see Rose et al. 2002, 30–32 for discussion of 'conception Dreamings'). Setting human reproduction aside, it is clear from a variety of other contexts that the bringing forth of life into the world is work done by men and women. It is work that engages with the life which is contained within sites or sources, and it seeks to bring life forth from the site of containment to active life in the world. I understand it to be work that triangulates across women, men and Country, bringing contained life into permeable and ephemeral bodies, and sustaining the relationships of connection.

Baby spirits

Dreaming Women left unborn babies at various sites in their travels. These are the repositories of future generations. They are contained in the Country, they are for the Country, and new generations of people are born by being shifted from a baby spirit site out into its mother's womb to be nurtured by her blood and later to be born onto the ground in the mother's blood.

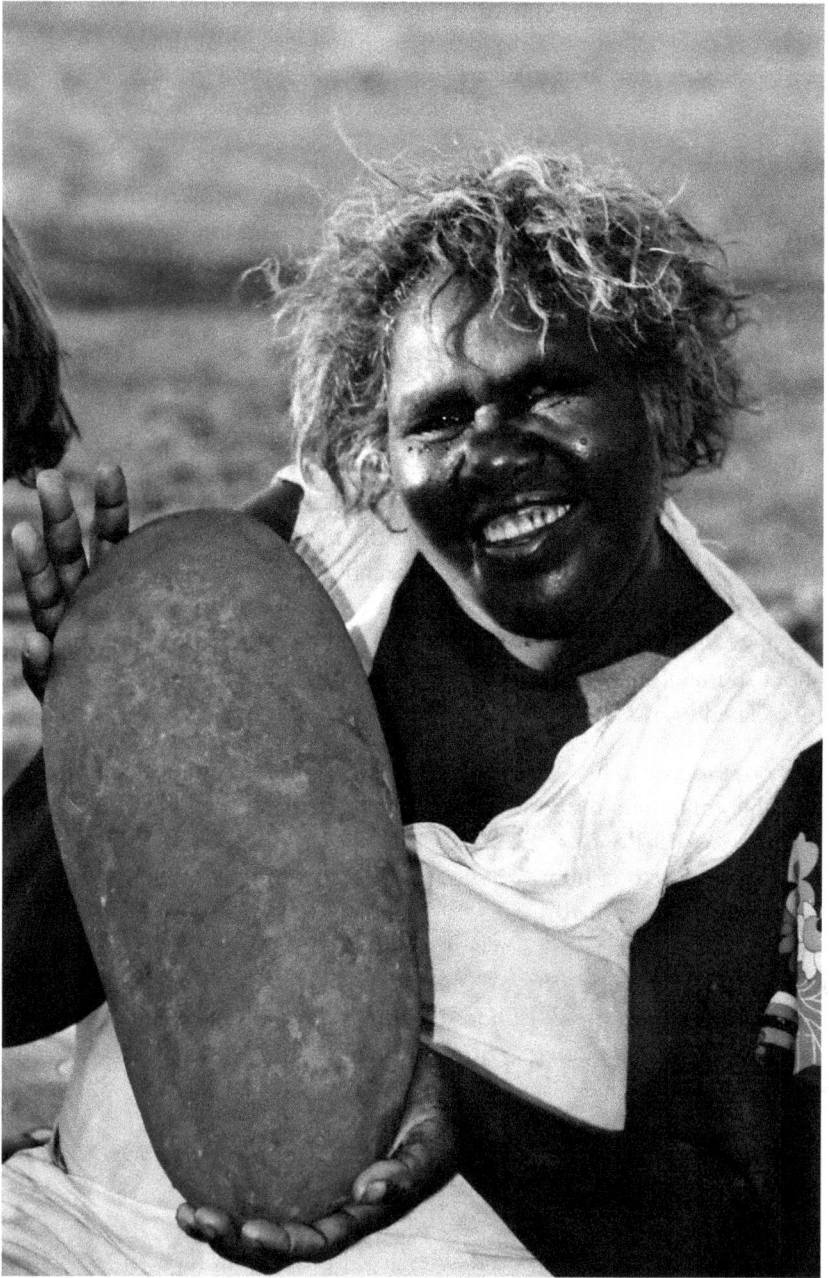

Figure 9.1. Ivy Kulngarri holding a Baby Spirit Dreaming stone, Pigeon Hole, 1986.

Source: Photograph by Darrell Lewis.

Stories about how the shift from a baby source to a mother's womb is effected vary, as, indeed, ideas about the number and sources of animating spirits may vary to some degree (see Rose 1992, 58). In general, however, an unborn baby who is looking for birth will leap into an animal. The father kills the animal and gives food to the mother, the mother eats, and through one or another sign (most frequently vomiting) the mother comes to know that a baby spirit has inhabited her. The transition of spirits from earthly source to motherly source is thus mediated by animals and effected through hunting and eating. Chains of connection give the person a genealogy that not only includes parents, but also Country, Dreaming source and the mediating animal, as well as embedding the person's very existence in relationships of connection, vulnerability and death.

Hobbles's genealogy is a good example. He was a barramundi before he became a person. His father speared the fish, his mother ate some, and the spirit became the baby who grew into the man known as Hobbles Danaiyarri. On his right temple he had a small mark where his father speared the fish. Hobbles was born well within the period of colonisation, and his genealogy demonstrated that. In the early days a group of Aboriginal people had been fishing somewhere in the Wave Hill area and were shot by white fellows. One of the men died in the water. His spirit became a barramundi; the barramundi became Hobbles.

Permeable bodies

The Dreaming Women who came walking across the land were shapeshifters. In their woman form they carried babies and gave birth, and those places remain as sites of origin for the people and the Law of birth and belonging that goes with that land. In other places they menstruated, and the blood remains in the ground today. Such sites are dangerous: the blood of the Dreaming Women is powerful still today. Many of the blood sites are red ochre deposits. Some belong exclusively to the women of the Country where the site is located; some belong exclusively to the men. Both men and women have an interest in protecting the integrity of the blood of the Dreaming Women, and both men and women use that blood/ochre in bringing life and growth into the world.

Menstruation and menstrual blood are not suitable topics for public discussion, and even in private they may be dangerous. I do not delve beyond the surface of these issues, but I want to accomplish two purposes.

The first is to show some of the contrasts with western views of menstrual blood because the differences are instructive in thinking about life and death. The second is to show that blood works to bring forth life, and thus is valued as life is valued.

Feminist studies of gendered bodies in the western world have taught us to understand bodies as corporeal repositories of particular kinds of subjectivity. Thus, western liberal democracies hold a concept of subjecthood, or subjectivity, which is implicitly male, and which is incorporated individually. The individual subject is an individual body with clearly and cleanly demarcated boundaries. Julia Kristeva's concept of 'le corps propre' speaks to this cleanly and clearly bounded entity (discussed in detail in Rosengarten 1996, 16–27). Elizabeth Grosz makes the further point that the good subject is in control of a good body—that is, the body is fully enclosed, and is solid. In contrast to these images of the bodied subject, women's bodies appear to be excruciatingly transgressive (1994). The wetness, fluidity and the 'leakiness' of menstrual blood escapes the requirements of a clean and proper body. The fact that blood, unlike tears, urine and other substances, cannot be suppressed at will contributes to the out-of-control sense of women's improper bodies. The permeability alluded to by the leakage of blood suggests a lack of proper subjecthood, so that for women to achieve subject parity in western societies, menstrual blood must be hidden, and the fact of menstruation concealed from public knowledge (Rosengarten 1996, 27–66). Thus, as Marsha Rosengarten (1996, 33–37) shows, advertisements for sanitary napkins emphasise protection against the shame of blood escaping into sight. Their absorbency is promoted for preventing leakage and thus overcoming imagery of a wet and out-of-control body. Advertisements for tampons refigure the imagery further, emphasising a complete absence of blood exterior to the body, and thus offering liberation from the need to deal with boundaries. The tampon-imagined body is in control of its boundaries and in parity with male bodies.

I offer this summary of feminist analysis because it is so massively contradicted by what we learn from Dreaming Women and from Aboriginal action in respect of menstrual blood. I will track ideas about life and permeability in Victoria River culture before returning to further discussion of blood.

Connectivities

Connections go deep. They implicate people, other living things and places in the wellbeing of each other's lives. As I have earlier discussed, flesh is shared across species so that human life is implicated in the lives of other species; Country provides nourishment and is always at risk of flood, drought and other major forms of change. An event has consequences that ripple across connections, and the subjectivity of any given living thing is not confined by the boundaries of the skin, but rather is sited inside, on the surface of and beyond the body. Subjects are constructed both within and without; subjectivity is located within the site of the body, within the bodies of other people and other species, and within the world in trees, rock holes, on rock walls and so on. And of course, location is by no means random; Country is the matrix for the structured reproduction of subjectivities.

Several points of contrast stand out. First, for Victoria River people, permeability is a sign of connection rather than a sign of loss, and thus is the desired state of being. Second, the flow of blood from inside the person to outside the person generates powerful forms of connection. Third, blood is not something to be kept hidden; Dreaming Women's uterine blood is located in the world and is named (often elliptically) and visible, and women's menstruation is subject to knowledge and management. Fourth, bodies are not bounded in all contexts; the opening up of the body to the release of blood, or for childbirth, or in other (ritual) contexts is a key process in bringing life forth into connection.

As a substance, menstrual blood is exceedingly powerful. It is a gift from the Dreaming. It is powerfully related to life, and also to love and to lust. Menstrual blood incites men to desire, and young men are particularly cautioned to stay away from women and to have no contact with women's blood. Improperly managed, blood makes men 'wild'. Blood is dangerous because it is powerful. In the 1940s Catherine Berndt visited Aboriginal settlements at Wave Hill and a number of other cattle stations south and south-west of Victoria River Downs. She formed the view that women's blood is sacred, and while I will dispute the sacred/profane dichotomy (below), I believe that her statement accurately expresses the quality of the power that pervades anything to do with blood (Berndt 1950).

As an event, menstrual blood is an announcement: something is happening. The question is, what does menstrual blood announce? Cross-cultural anthropological studies of menstruation, meanings and taboos, indicate that blood may be seen as neutral, it may be understood as a sign of failed pregnancy and thus is associated with death, or it may be understood as a sign of life, and thus is associated with fertility (Rosengarten 2000). My evidence shows without doubt that in the Victoria River region menstrual blood is a sign of fertility. It announces that pregnancy may occur.

Recent studies in the frequency and regularity of menstruation suggest that Aboriginal women formerly would have experienced very few menstrual periods in the course of their lives. Later onset of menstruation (compared to the present) combined with long periods of lactation, and in all probability with periods during which body fat was below the critical threshold for menstruation, would have ensured that there would be long spells between periods. Diane Bell has stated that in her interviews with senior Aboriginal women, she found that women could remember each menstrual period of their fertile lives and count them on their fingers (quoted in Buckley and Gottlieb 1988, 45). Comparative perspectives suggest critical fat levels would have been most reliably sustained in resource-rich regions (generally coastal), and least reliably sustained in the demanding desert regions (see Cowlishaw 1981; Rosengarten 1996). In the Victoria River region, there was a period of abundance (late wet, early dry) and a period of stress (late dry, early wet, see Chapter 1). It is probable that amenorrhoea (absence of menstruation) would have been most common during stress times (especially during droughts), and that women who were ready to menstruate would be most likely to do so in the period shortly after the wet season when food was in abundance. This was also a period when people gathered for ceremony, and as most ceremonies concern fertility their efficacy may have been confirmed in part by the onset of menstruation amongst appropriate women, followed by pregnancies in due course.

Childbirth blood

In the recent past, before childbirth was medicalised, women gave birth onto the ground. The umbilical cord is a profound sign of the connection between mother and child, as is the placenta. Both were handled with care, according to Dreaming Law instituted for women in childbirth. The blood itself flowed into the ground and became a powerful form of connection between the person, the mother and the site of the blood. Riley Young spoke of these matters, and because he is a man it is clear that he is speaking from general (shared) knowledge. He links the Earth as mother with his own mother and his own birth, and links both with rituals that women perform in camp for newborn babies (mentioned in relation to various plants in Chapter 7). He pulls these different strands of birth and blood, mother and ground, together as Law (see Rose 1992, 61–65):

> Because this ground belong to Aboriginal people. Aboriginal people been born la this ground. What they call it this ground, he's [she's] the mother. He's the mother. We used to be born la this ground. No hospital, no needle, no medicine. Used to be working by cooking by ashes. Make this. Any boy been born la this ground, him [mother] bin makem strong by cooking by [rubbing with] ashes. Make it [that baby] strong. Because this ground is the hospital. Even me, [I] been born la this ground. My mother been used to rub me with the ground, makem me black one, same as that coal. Used to get that coal, grindem up. Or get a pandanus, or spinifex. Put em in and roastem up in the ground [rub the ashes on me], making me strong. That's what they been do: this Law, la this place. Early days. Because I never been born by top of the hospital. I been born by ground. Because I know this Law.

I will not pursue the discussion of the rituals for newborn babies, as that is discussed in detail elsewhere (Rose 1992, 61–65). Rather, I will move on to discuss birth as a paradigmatic metamorphosis by which life comes from the state of being contained within to the state of being on the ground outside the mother (or site of containment).

Metamorphosis

Woman Dreamings/Dreaming Women: Aboriginal Australia comes into being from the generative woman/body. From inside to outside, from life to life, from birth to death and back to birth, there is no transformation that does not emerge from and rely upon Dreaming Women and their human descendants, female and male. For Victoria River people, women's blood, the blood of menstruation and childbirth, is integral to the process of generating life. In talking about women's blood, especially the blood of the Dreaming Women, I am entering dangerous territory. The main point I want to make here is that in its sacred and powerful substance, women and men both manage it.

Much of what I know about how men manage women's blood derives from the category of information which is publicly available but never subjected to public exegesis. It consists of hints, clues, indirections, metaphors and sometimes even culturally consistent misdirections. Rather than going through these clues to construct an image of the knowledge and practice that lie behind them, I will develop an analysis out of published material that speaks to the pertinent issues. I take a short cut and quote from Lloyd Warner, an American anthropologist who worked with Arnhem Land men in the 1920s. He quoted an explanation offered by one of his Aboriginal teachers:

> That blood we put all over those men is all the same as the blood that came from that old woman's vagina [referring to a creative female Dreaming being]. It isn't the blood of those men any more because it has been sung over and made strong. The hole in the man's arm isn't that hole any more. It is all the same as the vagina of that old woman that had the blood coming out of it … When a man has got blood on him … he is all the same as those two old women when they had blood. (1958 [1937], 268)[3]

In the Victoria River District, subincision is the mark of a mature man, and the blood is let from the penis as well as the arm. It bears a value similar to what Warner explained. Warlpiri men whose homelands are out in the desert proper perform ceremonies which are restricted to men only, and which have the intention of invigorating the fertility of the Country (both in reference to specific totemic species, and more broadly in reference to overall fertility). In some ceremonies, certain men dance a dance that is similar to women's dancing (Wild 1977–78). The practice in one such dance is to reopen the subincision wound on the penis, so that as they dance their blood splatters their thighs and flows down their legs (19). Some authors have contended that this use of women's blood by men is a symbolic appropriation of women's reproductive power (Wild 1977–78, 1984; Swain 1993). I think this is misleading. Ideas of symbolic appropriation set men and women in opposition whereas they are actually in complementary and procreative relationships. More profoundly, however, it sustains, and even reinforces the idea that gender categories are given in nature. Culture, in these explanations, is a mechanism whereby the naturally given can be culturally supplemented, or even supplanted.

---

3   I quote Warner for the quality of his teacher's explanation, and not for his own analysis which, in this context, is heavily grounded in a sacred/profane dichotomy and fails to do justice to the multiplicity of sacred substances, beings and events.

With a more subtle understanding of the complexities of gender, we can see that ritual as a performative situates men as mothers so as to engage in the procreative rituals which bring forth the life of the world. Wild states that 'in men's rituals, the … men for whom the country is mother's country are substitute women; on occasions when they dance they perform in women's … styles' (1977–78, 18). It thus becomes clear that the blood from the male 'mother' is a flow of fertility making male menstrual blood. In any given ceremony, then, the men who dance for their mother's Country dance as 'mothers' and the men who dance for their father's Country dance as 'fathers'. Together, 'mother' and 'father', they dance into being the fertility of the Country.

Men's accounts of how they came to have a Law that originates with women emphasise both the competitive quality of a perceived lack of symmetry, and also the shared project of bringing forth new life. To emphasise one aspect (competition owing to asymmetry) and ignore another aspect (shared responsibility for new life) is to pull the Aboriginal evidence back into a domain of envy and appropriation that apparently is more congenial to western (male) thought.

An alternative way of engaging with men's and women's use of blood and the vehicles of birth in ritual contexts is to consider the co-substantiality of brother and sister. The effect of men dancing as women is to produce an equivalence between brother and sister (not husband and wife), but this is not subject to public verification because it is accomplished in gender-restricted domains. The published literature contains numerous statements that ensure the understanding that when men let flow their own blood in ritual, they are releasing the powerful blood of the Dreaming Women (Warner 1958 [1937], 268; Wild 1977–78).

I will close this dangerous topic with the conclusion that there is blood, and then there is blood. In ritual and in creation, women's uterine blood and its cultural equivalents such as arm blood, penis blood or red ochre is powerful and transformative. Who uses that blood, and what transformations they effect with it, is deep and secret business. The process of bringing the world into being is the work of women and of men, and it is never finished.

Desire

The Earth itself, both in its locatedness of Country and by extension more generally as the basis of the life in the whole world, is female. Everything comes out of the Earth, and in the beginning of creation all the Dreamings came out of the Earth; the holes or caves of emergence are wombs. Earth is mother, living things are all connected to the Earth, and by implication to each other in some unspecified network of kinship.

Somewhat to my surprise, at least one scholar has interpreted my writings, and some statements by Riley Young, as evidence both for my desire to find an Earth Mother, and Riley's invention of a cultural form in which to express some of the pain of invasion (Swain 1993, 195–96). I do not want to belabour an argument that deflects me from what I take to be the serious issues here, but I also want to be sure that I am clear in what I am saying. Tony Swain claims that in *Dingo Makes Us Human* I offer only one 'very brief piece of ethnographic data referring to "Earth as Mother"' (his capitalisations), and his opinion is that this one example is 'quite evidently a recent post-colonial creation' (1993, 196). He also suggests that I draw on Gaia 'to illuminate [the] Aboriginal data' and that this proves the circularity of academic traditions. Swain evidently missed portions of my book that quite explicitly refer to the origin of things within the Earth and their emergence to the surface of the Earth. I had intended to be clear, if sometimes elliptical, in equating this process of emergence with the process of birth. My reference to Gaia was not a reference to the Greek goddess but rather to James Lovelock's Gaia hypothesis—named for the Greek goddess, but actually defining a very recent move in ecological thinking that treats the planet (including the atmosphere) as a self-regulating organism (Bouissac 1998; Lovelock 1979). The reason for citing Lovelock was to identify a shift in western thought, not to bolster Indigenous thought.

The actual statement that Swain contends proves my complicity in 'Earth Mother', or as he also phrases it, 'All-Mother', thinking is a quote from my teacher Riley Young: 'This ground she's my mother. She's mother for everybody. We born top of this ground. This our mother. That's why we worry about this ground' (Swain 1993, 197; see also Rose 1992, 220). Clearly, Riley Young is not talking about an Earth Mother goddess or an All-Mother. He is talking about kinship. From kinship to a common mother, he asserts the basis of claiming kinship with other people and other living things. And from these kinship connections he asserts a community of care. In other statements he wonders why the necessity for a community of care is not perceived by white fellows.

Another prong of Swain's analysis is that certain of the Dreaming figures are the product of contact with Macassar and other foreign visitors. He believes that they articulate new social and land issues that arise out of contact with strangers, and that they are ultimately traceable to the Rice Goddess of Southeast Asia. In his view, these Dreaming figures are Earth Mother goddesses, modified by Aboriginal thinkers to articulate an Aboriginal ontology.

Many aspects of Swain's analysis are arguable and have been and will continue to be argued (Keen 1993; Rose 2000). Ian Keen, an anthropologist with long and deep experience in Arnhem Land, exposes many of the flaws in the argument both in relation to Arnhem Land and more generally. He notes, for example, that Dreaming tracks of creator women are so widespread across Australia that they are unlikely to have come into being in the recent past. I do not have ethnographic expertise in Arnhem Land, and I have no opinion on the possible influences of the Southeast Asian Rice Goddess. Swain quite correctly identifies a new religious movement focusing on female fertility, and it is well established in the Victoria River region, and among the Warlpiri people to the south who have over the years shifted emphasis from male to female (Wild 1977–78) in the ceremony. Does this mean it had to come from Asia? Nothing in my work suggests that the Dreaming Women are a new appearance in the world, and nothing suggests that female fertility is a novelty that had to be stimulated by contact with foreigners. It seems perverse to imagine that for some 60,000 years Aboriginal people ignored fertility, and that then in the past few centuries they entered a period of cultural innovation in which they took up concepts of fertility (including female fertility) as a key problematic, and as root metaphors for life in the ephemeral world.

Swain also argues that some recent statements by Aboriginal people concerning Earth as mother actually reference a recent idea—that of Earth as the colonised and defenceless Mother. His desire to cast Riley's statement in a mode of defenceless Mother is absolutely untenable. In Riley's view, the ground (which one might gloss both as 'nature' or as 'Earth') is a non-negotiable force in the world. He considers, for example, government plans to alter the course of rivers:

> Why that government reckon he gonna changem everything? Change him round? How you going to change him round? You can't change … that big hill there. You can't change him this ground. How you going to change him? How you going to change that creek? … Put that creek this side, he'll come back to flood this side. You can't! No way!
>
> I know government say he can change him rule. But he'll never get out of this ground. (Quoted in Rose 1992, 57)

There is no defenceless and pathetic All-Mother in this statement. There is a vigorous critique of 'government', and of the idea that it can control the Earth.

~ ~ ~

Returning to my main purpose here, Earth was the originator of life in a kind of completion that is not procreative as I am using the term here. Earth existed and brought forth Dreamings without other agency. The Dreamings walked around creating diversity, connectivity and pattern.

In thinking about the stories of Dreaming origins, I contemplate the relationship between static power and power in motion. The ordinary Earth holds the power contained, and origin stories tell us that the life contained is life in some sense unfulfilled. Creation overcomes this lack: the Earth gave birth to the Dreamings, the Dreamings came forth and walked the Earth, giving shape, boundaries, connections, Law.

I believe it is reasonable to say that the power contained within the Earth has desire. That desire is to be embodied and mobile. Life wants to go walking around on the surface; it wants to live in the ephemeral world of bodies and motion, as well as in the inside world of containment. Furthermore, it desires pattern and connection; it wants to flourish. Life thus exists as both the enduring potential contained within and a dynamic and flourishing ephemeral actuality that lives and dies on the surface.

Earth contained the life within, and Earth brought it forth. Today, however, things are different: the continuing bringing forth of life requires a plurality of agency. The relationships are between Earth as original and containing source and the ephemeral life that now works in procreative action to bring forth more life. I have discussed fire as one of the major forms of action that promotes life. I now turn to ceremony as another context within which primary life paradigms are expressed and renewed, in which life's desire to be mobile is reconciled with life's desire to be contained.

Ceremony

Many ethnomusicologists and anthropologists have found that Aboriginal people assert that ceremonies work to bring Dreaming power and presence into the time, place and bodies of the performers (see Ellis 1985, 94). Different scholars use a range of terms for talking about this process, but Cath Ellis, more than many, seeks to unpack the performative elements that make possible this coming together of ephemeral human life and the world creative Dreamings.

Ellis asserts that time is a crucial technique in music, and she analyses the complex interlocking of a multiplicity of patterned elements:

> From the smallest element of the fixed duration of the short notes setting the song text, through the beating duration, the repeated rhythmic segments, rhythmic patterns, text presentations, to melody, small song and songlines—each using its own time-scale, and each a series of intermittently emphasised patterns such that first one, then another occupies the centre of attention—a ceremony … is unfolded. (1984, 160)[4]

Elements are nested within broader elements, according to Ellis, and each element has its own time. The genius of Aboriginal performers is to interlock these multiple patterns into a whole pattern. Ellis explains:

> The total pattern, apparently known in advance, is divided accurately and proportionately, at points of no other musical significance, even when the pattern does not commence at the beginning of a line of the song text. These divisions, therefore, occur in advance of the total pattern having been presented. (1984, 163)

There are patterns within patterns, all of which have to mesh. Aboriginal people, far from being a people without time as some scholars have suggested, have the time equivalent of the gift we know as perfect pitch. Some Aboriginal performers, Ellis contends, are masters at organising

---

4   I am not attempting to deal with Ellis's discussion of time, which I have previously discussed in Rose (2000).

patterns in advance so that mathematically complex divisions fall correctly into the total pattern. This is done without reference to mathematics, and thus depends on some other faculty which Ellis (1984, 184) labels 'perfect time'.

In Ellis's view (1985, 109), the correct interlocking of all the nested and coexisting patterns generates the strength by which song draws the power of the Dreamings out of the Earth. Moments of perfect pattern constitute cosmogonic action that lifts the Dreamings up from the Earth and enables them to become mobile, being carried by the participants. This is serious business. As Ellis says, to lose the power of Dreaming or ancestral connection in ceremony is 'unforgivable' (1985, 109).

Ceremony is performative—it brings the Dreamings up out of the ground and carries them through the Country. Ellis's work is directed toward song; here I expand her analysis and link it to dance and other aspects of ceremony. In the part of the Victoria River valley where I have danced on many occasions, the ceremony is called *Pantimi*. The songs sing the track of a group of creative Dreaming Women. The men sing the songs and the women dance. I learned to dance, and so I learned to work the ground with my feet and learned to make the dance-call that is integral to the pattern. Thus, I learned that the body connects Earth and air when you dance. The call comes from deep within and is propelled by the impact of your feet on the ground. It comes to feel as if the ground itself propels your voice out into the night sky. That call starts somewhere below your feet and ends somewhere out in the world. The call is a motion, a sound, a wave of connection. You are dancing the Earth, and the Earth is dancing you, and so perhaps you are motion, a sound, a wave of connection. You are a bearer of the call, and perhaps you are also a bearer of an answer.

I draw on Ellis's assertion that Aboriginal music is 'iridescent'. She explains this unexpected concept with reference to the phenomenon that occurs when background and foreground suddenly flip. Everyone experiences this phenomenon in visual form, particularly with art or photos that are designed to generate the flip between background and foreground. The flip phenomenon is also experienced aurally, as one or another pattern is heard as foreground. The song becomes iridescent through the complexities of the shifting ground of interweaving patterns.

Figure 9.2. Yarralin women, Debbie and her daughter Chantal dancing *Pantimi*, Yarralin, 1981.

Source: Photograph by Darrell Lewis.

For the dancer there are also embodied iridescences. There is the flip between the feet on the ground and the ground on the feet: who is the dancer and who is the danced? If I hold the analytic privilege on motion, I find that both are dancer and danced, and that the significance of this mutuality is located in the flip back and forth. In dance one is always engaged and always in motion, and in the moment when all is iridescent there is an exultant awareness of life in action.

Each segment of song and dance, however, is set apart by a counterpoint of non-dance. Each small song is punctuated by a pause, a break in the music. The rhythms of the song and dance are thus set within a larger oscillation of music and non-music. Ellis (1984, 170) notes that the pattern of the music is retained perfectly across the breaks within which there is no music, and the breaks are dedicated to joking. It is not a break in the ceremony but rather a contrapuntal engagement with the musical portion of the ceremony.

In the Victoria River District, there are some set joking themes, particularly the theme of women beating their mother's brothers, and uncles swearing at their nieces. The joking runs concurrently through the intervals, carrying themes of gender, sexuality, authority and spontaneous inventive delight.

Ceremony thus works with two interwoven event types: the music and dance are Dreaming Law and are internally and complexly patterned; the joking is spontaneous. Each joking interval is a qualitative and purposeful withdrawal from the song. Each song is a qualitative and purposeful return.

It would not be accurate to privilege either the musical performance or the joking. Nor would it be accurate to subsume one within the other. Rather, analytic privilege belongs in the movement back and forth between musical performance and joking performance and includes the dance/non-dance movement as another form of iridescence.

In *Pantimi* ceremonies, joking speaks of the ephemeral; of the spontaneous, the partial, the incomplete and the contingent. Performance engages Dreaming power as it is contained within the Earth; the call is performed in patterns that already are given, are intensely rule-governed and require proper execution. Joking and music call participants back and forth between the ephemeral and the enduring, until each can be seen to be implicated in the dance of the other. The ephemeral draws close to, and withdraws from, enduring creation power which itself approaches and withdraws. This motion captures a mutual embeddedness of the ephemeral in the enduring and the enduring in the ephemeral.

In ceremony, one becomes part of the pattern, and to become part of the pattern is to join in the call. In the between place of iridescence, a further question arises: Who is calling and who is called? To become part of the call is also (when things go properly) to become part of the response. One is transformed from agent (calling) to vehicle (being called or moved through) and back and forth all night long.

To dance, therefore, is to move within a generative, liminal matrix of betweens—between the caller and the respondent, between the ground and the foot, the Earth and the air; between the many interlocking patterns and flips, and between the enduring and the ephemeral. In the transformative between, one becomes connection. Life's desire flows through the Country and the person in waves and patterns of connection.

'Holding'

Allan Young spoke of 'holding' the Country when he said that his people never burnt the Country (with the implication that they never burnt it out); their work was to hold the Country. His words are like those of many other Aboriginal people (for example, see Mussolini Harvey in Bradley 1988, x–xii; see also Bell 2002, 265).

Ceremonies are episodic, and between the episodes are the periods of daily life (non-ceremony). In daily life Country holds the lives of its living things. People hunt and gather, animals and plants nourish themselves and each other, seasons shift, living things respond, the ephemeral world lives and dies. The alternative moment is ceremony. In the interplay between ceremony and daily life, creation nourishes the ephemeral and, episodically, the ephemeral calls up the enduring and brings it into the Country in new waves of patterned life. Daily life, one therefore would have to say, is not an interruption of life's dance, but is rather the counterpoint to the ceremonial part of the dance. It would not be appropriate to privilege either. Daily and ceremonial life, and flips between them, generate their own iridescences. The movement between is the continuous becoming of the created world.

The second point follows from the first. When people paint the designs on their bodies, they are carrying sites and species into the arena of ceremony. In this way, places within their Country such as rock holes, hills and specific landforms are brought into the full experience I have described for a person. So, too, are all the plants and animals that are carried on the body in the form of marks. Just as the person experiences the iridescence of ceremony, so too do the places and other living things brought into ceremony. Just as the person flips back and forth, enhancing and revitalising connectivities while being enhanced and revitalised by them, so too do places and other living things. And just as the person becomes both agent and vehicle, so too do places and other living things.

Life itself thus comes into being in between. It cannot be located wholly in the enduring potential, or in the ephemeral actual, but arises in the dance between them. And so again we have to ask who is dancing? It seems that life's desire dances with life's forms and patterns, and that the dance becomes the flow and form of life itself.

Immanence/'holding'

The great Arnhem Land sage David Burrumarra explained many complex things in language that is so suggestive as to be always open to more engagement. One can read his words and ponder them, have conversations with them and dream about them, but one can never fully fathom the depths of his insight. He tells us that *motj* (a term that is variously translated as power, spirit, the sacred) is power, that it is the source of all life. In comparing his sense of the sacred with Christianity, Burrumarra said:

> The Bible and the Cross help us to remember Christianity and to believe in God … They are like eyeglasses. Without these glasses would we see God in our image (and vice versa) or would God look different? Would he look like the natural world? (Burrumarra with Ian McIntosh 2002, 10)

I understand Burrumarra to imply that the answer to his last question is 'yes'. Power, without the metaphysical glasses of Christianity, is visible in the living world around us. I link this metaphysics to desire. Power desires its own becoming; it wants to enter the world of transience; it loves life and wants to live. This world of transience, passion and joy is the form power takes on the outside. Origin myths tell us this, and we learn this also in ceremony.

Appendix 1.
Letter from Brian Pedwell

Victoria Daly

REGIONAL COUNCIL
Regional Office 29 Crawford Street,
KATHERINE NT 0850

PO Box 19
KATHERINE NT 0851

Telephone 08 8972 0777
Facsimile 08 8971 0856
admin@vicdaly.nt.gov.au
www.victoriadaly.nt.gov.au

12 April 2023

Darrell Lewis
darrelllewis66@gmail.com

Dear Mr. Lewis,

RE: Photos for Dreaming Ecology

Thank you for reaching out to me regarding the images for the book 'Dreaming Ecology'. Following our discussions, I travelled to meet with members of the families to which the photos belong, and they were ecstatic to see them and enjoyed reminiscing on times past. They expressed their consent for the use of the photos in the book inclusive of all photos that I have received via email from you to date.

Please don't hesitate to contact me on 0429 341 336 or at Brian.Pedwell@vicdaly.nt.gov.au if you require any additional assistance.

Yours faithfully,

Brian Pedwell
MAYOR - WALANGERI WARD
0429 341 336 brian.pedwell@vicdaly.nt.gov.au
VICTORIA DALY REGIONAL COUNCIL

MOVING FORWARD TOGETHER

Bibliography

Primary sources: Debbie's teachers

All tapes are stored with Libraries and Archives Northern Territory, PCA 11, Dr Deborah Bird Rose Personal Collection.

Tapes

Allan Young, tape 56, recorded at Yarralin, no date recorded.

Allan Young, tape 116, recorded by Darrell Lewis at Katherine, 24 August 2000.

Big Mick Kangkinang, tape 78, recorded at Yarralin, 13 July 1986.

Billy Bunter, tapes 114 and 115, recorded by Darrell Lewis at Daguragu, 19 August 2000.

Charcoal Winpara, tape 84, recorded at Yarralin, 21 July 1986.

Charcoal Winpara and Jambo Muntiyari, tape 92, recorded at Yarralin, 31 July 1986.

Daly Pulkara, tape 80, recorded at Lingara, 15 July 1986.

Dora Jilpngarri, tape 74, recorded at Yarralin, 15 April 1988.

Dora Jilpngarri, tape 82, recorded at Yarralin, 18 July 1986.

Doug Campbell, tape 86, recorded at Yarralin, 24 July 1986.

Doug Campbell, tape 87, recorded at Yarralin, 25 July 1986.

Hobbles Danaiyarri, tape 1, recorded at Yarralin, October 1980.

Hobbles Danaiyarri, tape 16, recorded at Lingara, 11 April 1982.

Hobbles Danaiyarri, tape 91, recorded at Pigeon Hole, 27 July 1986.

Jessie Wirrpa, tape 78, recorded at Yarralin, 13 July 1986.

Jessie Wirrpa, tape 79, recorded at Yarralin, 15 July 1986.

Old Jimmy Manngayarri, tape 109, recorded at Yarralin, 13 August 1991.

Old Jimmy Manngayarri, tape 110, recorded at Yarralin, 13–14 August 1991.

Old Jimmy Manngayarri, tape 111, recorded at Daguragu, 14 August 1991.

Old Jimmy Manngayarri, tape 114, recorded by Darrell Lewis at Daguragu, 19 August 2000.

Kitty Lariyari, tapes 85 and 86, recorded at Yarralin, 24 July 1986.

Nancy Kurung, tape 29, recorded at Yarralin, 12 April 1982.

Riley Young, tape 42, recording date and location unknown.

Riley Young, tape 86, recorded at Yarralin, 24 July 1986.

Snowy Kulmilya, tapes 89 and 90, recorded at Yarralin, 27 July 1986.

Notebooks

Big Mick Kangkinang, notebooks 12, 20, 21 and 39.

Daly Pulkara, notebooks 3 and 37.

Dora Jilpngarri, notebooks 6, 52 and 53.

Doug Campbell, notebook 38.

Hobbles Danaiyarri, notebooks 11, 13 and 17.

Old Jimmy Manngayarri, notebook 55.

Nancy Kurung, notebook 18.

Riley Young, notebook 55.

Old Tim Yilngayarri, notebooks 2 and 4.

Secondary sources: Books, journals, etc.

Ankersmit, Franklin R. 1994. *History and Tropology: The Rise and Fall of Metaphor.* Berkeley, CA: University of California Press. doi.org/10.1525/9780520309814.

Aplin, Graeme J, Paul Beggs, Gary Brierley, Helen Cleugh, Peter Curson, Peter Mitchell, Andrew Pitman and David Rich. 1996. *Global Environmental Crises: An Australian Perspective*. Melbourne: Oxford University Press.

Arthur, Jay Mary. 1996. *Aboriginal English: A Cultural Study*. Melbourne: Oxford University Press.

Athanasiou, Tom. 1997. *Slow Reckoning: The Ecology of a Divided Planet*. London: Secker & Warburg.

Australian Environment Council. 1982. *Report on Litter Control*. AEC Report No. 8. Canberra: Australian Government Publishing Service.

Bateson, Gregory. 1972. *Steps to an Ecology of Mind: Collected Essays in Anthropology, Psychiatry, Evolution, and Epistemology*. San Francisco, CA: Chandler Pub. Co.

Beale, Bob, and Peter Fray. 1990. *The Vanishing Continent: Australia's Degraded Environment*. Sydney: Hodder & Stoughton.

Bell, Diane. 1983. *Daughters of the Dreaming*. Melbourne: McPhee Gribble; Sydney: Allen and Unwin. (Third edition 2002).

Benterrak, Krim, Stephen Muecke and Paddy Roe, with Ray Keogh, Butcher Joe (Nangan) and EM Lohe. 1984. *Reading the Country: Introduction to Nomadology*. Fremantle: Fremantle Arts Centre Press.

Berndt, Catherine. 1950. 'Expressions of Grief among Aboriginal Women'. *Oceania* 20 (4): 286–332. doi.org/10.1002/j.1834-4461.1950.tb00166.x.

Berndt, Ronald M, and Catherine H Berndt. 1987. *End of an Era: Aboriginal Labour in the Northern Territory*. Canberra: Australian Institute of Aboriginal Studies.

Bird-David, Nurit. 1990. 'The Giving Environment: Another Perspective on the Economic System of Gatherer-Hunters'. *Current Anthropology* 31 (2): 189–96. doi.org/10.1086/203825.

Bird-David, Nurit. 1992a. 'Beyond "the Hunting and Gathering Mode of Subsistence": Culture-Sensitive Observations on the Nayaka and Other Modern Hunter-Gatherers'. *Man*, New Series (renamed *Journal of the Royal Anthropological Institute*) 27 (1): 19–44. doi.org/10.2307/2803593.

Bird-David, Nurit. 1992b. 'Beyond "the Original Affluent Society": A Culturalist Reformulation [and Comments and Reply]'. *Current Anthropology* 33 (1): 25–47. doi.org/10.1086/204029.

Bird-David, Nurit. 1993. 'Tribal Metaphorization of Human–Nature Relatedness: A Comparative Analysis'. In *Environmentalism: The View from Anthropology*, edited by Kay Milton, 112–25. Association of Social Anthropologists Monograph Series. London: Routledge. doi.org/10.4324/9780203449653_chapter_8.

Birdsell, Joseph B. 1953. 'Some Environmental and Cultural Factors Influencing the Structuring of Australian Aboriginal Populations'. *American Naturalist* 87: 171–207. doi.org/10.1086/281776.

Birdsell, Joseph B. 1968. 'Some Predictions for the Pleistocene Based on Equilibrium Systems among Recent Hunter-Gatherers'. In *Man the Hunter*, edited by Richard B Lee and Irven DeVore, 229–40. Chicago: Aldine Publishing Company. doi.org/10.4324/9780203786567-29.

Bolton, Geoffrey. 1981. *Spoils and Spoilers: Australians Make Their Environment 1788–1980*. Sydney: George Allen & Unwin.

Bouissac, Paul, ed. 1998. *Encyclopaedia of Semiotics*. Oxford: Oxford University Press.

Bowman, David. 1995. 'Why the Skillful Use of Fire Is Critical for the Management of Biodiversity in Northern Australia'. In *Country in Flames: Proceedings of the 1994 Symposium on Biodiversity and Fire in North Australia*, edited by Deborah Bird Rose, 103–10. Biodiversity Series 3. Canberra: Biodiversity Unit, Department of the Environment, Sport and Territories; Darwin: North Australia Research Unit, ANU. Online: openresearch-repository.anu.edu.au/bitstream/1885/282735/1/b18944231.pdf, accessed 31 January 2023.

Bowman, David MJS, and William J Panton. 1993. 'Decline of *Callitris intratropica* R.T. Baker & H.G. Smith in the Northern Territory: Implications for Pre- and Post-European Colonization Fire Regimes'. *Journal of Biogeography* 20: 373–81. doi.org/10.2307/2845586.

Boyarin, Daniel. 1990. 'The Eye in the Torah: Ocular Desire in Midrashic Hermeneutic'. *Critical Inquiry* 16 (3): 532–50. doi.org/10.1086/448545.

Bradley, John. 1988. *Yanyuwa Country: The Yanyuwa People of Borroloola Tell the History of Their Land*. Richmond, VIC: Greenhouse Publications.

Bradley, John. 1995. 'Fire: Emotion and Politics; A Yanyuwa Case Study'. In *Country in Flames; Proceedings of the 1994 Symposium on Biodiversity and Fire in North Australia*, edited by Deborah Bird Rose, 25–31. Biodiversity Series 3. Canberra: Biodiversity Unit, Department of the Environment, Sport and Territories; Darwin: The North Australia Research Unit, ANU. Online: openresearch-repository.anu.edu.au/bitstream/1885/282735/1/b18944231.pdf, accessed 31 January 2023.

Braiterman, Zachary. 1998: *(God) After Auschwitz: Tradition and Change in Post-Holocaust Jewish Thought.* Princeton, NJ: Princeton University Press. doi.org/10.1515/9781400822768.

Braithwaite, Richard W. 1995. 'A Healthy Savanna, Endangered Mammals and Aboriginal Burning'. In *Country in Flames; Proceedings of the 1994 Symposium on Biodiversity and Fire in North Australia,* edited by Deborah Bird Rose, 91–102. Biodiversity Series 3. Canberra: Biodiversity Unit, Department of the Environment, Sport and Territories; Darwin: The North Australia Research Unit, ANU. Online: openresearch-repository.anu.edu.au/bitstream/1885/282735/1/b18944231.pdf, accessed 31 January 2023.

Bright, April. 1995. 'Burn Grass'. In *Country in Flames; Proceedings of the 1994 Symposium on Biodiversity and Fire in North Australia,* edited by Deborah Bird Rose, 59–62. Biodiversity Series 3. Canberra: Biodiversity Unit, Department of the Environment, Sport and Territories; Darwin: The North Australia Research Unit, ANU. Online: openresearch-repository.anu.edu.au/bitstream/1885/282735/1/b18944231.pdf, accessed 31 January 2023.

Bringing Them Home: Report of the National Inquiry into the Separation of Aboriginal and Torres Strait Islander Children from Their Families, April 1997. 1997. Australian Human Rights Commission: Commonwealth of Australia. Online: humanrights.gov.au/our-work/bringing-them-home-report-1997, accessed 18 January 2023.

Brock, John. 1988. *Top End Native Plants.* Darwin: J. Brock.

Brock, Peggy, ed. 1989. *Women, Rites and Sites: Aboriginal Women's Cultural Knowledge.* London: Allen & Unwin.

Brody, Hugh. 1981. *Maps and Dreams: Indians and the British Columbia Frontier.* Vancouver: Douglas & McIntyre.

Buber, Martin. 1970 [1937]. *I and Thou.* Translated by W. Kaufmann. New York, NY: Charles Scribner and Sons.

Buckley, Thomas, and Alma Gottlieb. 1988. 'Introduction: A Critical Appraisal of Theories of Menstrual Symbolism'. In *Blood Magic: The Anthropology of Menstruation,* edited by Thomas Buckley and Alma Gottlieb, 3–50. Berkeley, CA: University of California Press. doi.org/10.1525/9780520340565.

Bureau of Meteorology. n.d. Online: www.bom.gov.au/climate/averages/tables/cw_014825.shtml, accessed 18 January 2023.

Burrumarra, David, with Ian McIntosh. 2002. 'Motj and the Nature of the Sacred'. *Cultural Survival Quarterly Magazine* 26 (2). Online: www.culturalsurvival.org/publications/cultural-survival-quarterly/motj-and-nature-sacred, accessed 18 January 2023.

Cheney, Jim. 1989. 'Postmodern Environmental Ethics: Ethics of Bioregional Narrative'. *Environmental Ethics* 11 (2): 117–34. doi.org/10.5840/enviroethics 198911231.

Cowlishaw, Gillian. 1981. 'The Determinants of Fertility among Australian Aborigines'. *Mankind* 13 (1): 37–55. doi.org/10.1111/j.1835-9310.1981.tb 01216.x.

Crawford, Ian. 1982. 'Traditional Aboriginal Plant Resources in the Kalumburu Area: Aspects in Ethno-Economics'. *Records of the Western Australian Museum,* Supplement No. 15.

Critchley, Simon. 1991. '"Bois"—Derrida's Final Word on Levinas'. In *Re-Reading Levinas,* edited by Robert Bernasconi and Simon Critchley, 162–89. Bloomington, IN: Indiana University Press. doi.org/10.5040/9781472547354.ch-010.

Curthoys, Ann, and John Docker. 1999. 'Time, Eternity, Truth and Death: History as Allegory'. *Humanities Research* 1: 5–26. doi.org/10.22459/HR.01.1999.01.

Davis, Richard. 2005. 'Eight Seconds: Style, Performance and Crisis in Aboriginal Rodeo'. In *Dislocating the Frontier: Essaying the Mystique of the Outback,* edited by Deborah Bird Rose and Richard Davis, 145–63. Canberra: ANU E Press.

Davis, Stephen. 1997. 'Documenting an Aboriginal Seasonal Calendar'. In *Windows on Meteorology: Australian Perspective,* edited by Eric K Webb, 29–33. Melbourne: CSIRO Publishing.

de Certeau, Michel. 1992. *The Mystic Fable.* Chicago, IL: Chicago University Press.

Dietrich, William. 1992. *The Final Forest: The Battle for the Last Great Trees of the Pacific Northwest.* New York, NY: Penguin Books.

Douglass, Paul. 1992. 'Deleuze and the Endurance of Bergson'. *Thought: Fordham University Quarterly* 67 (1): 47–61.

Ecologically Sustainable Development (ESD) Working Groups. 1991. *Final Reports.* 12 vols. Canberra: Australian Government Publishing Service. Online: www. vgls.vic.gov.au/client/en_AU/vgls/search/detailnonmodal/ent:$002f$002fSD _ILS$002f0$002fSD_ILS:55101/ada?qu=Sustainable+development.&d= ent%3A%2F%2FSD_ILS%2F0%2FSD_ILS%3A55101%7EILS%7E19& ps=300&h=8, accessed 13 September 2023.

Elkin, Adolphus Peter. 1933. *Studies in Australian Totemism.* Oceania Monographs 2. Canberra: Australian National Research Council.

Elkin, Adolphus Peter. 1954 [1938]. *The Australian Aborigines: How to Understand Them.* Sydney: Angus and Robertson.

Ellis, Catherine. 1984. 'Time Consciousness of Aboriginal Performers'. In *Problems and Solutions: Occasional Essays in Musicology, Presented to Alice Moyle*, edited by Jamie C Kassler and Jill Stubington, 149–85. Sydney: Hale & Iremonger.

Ellis, Catherine J. 1985. *Aboriginal Music: Education for Living: Cross-Cultural Experiences from South Australia*. St Lucia: University of Queensland Press.

Flannery, Tim. 1995. *The Future Eaters*. Chatswood, NSW: Reed Books.

Flores, Dan. 1998. 'Spirit of Place and the Value of Nature in the American West'. In *A Sense of the American West: An Anthology of Environmental History*, edited by James E Sherow, 31–38. Albuquerque, NM: University of New Mexico Press.

Frazer, James G. 1910. *Totemism and Exogamy: A Treatise on Certain Early Forms of Superstition and Society*. 4 vols. London: Macmillan.

Freud, Sigmund. 1919. *Totem and Taboo: Resemblances between the Psychic Lives of Savages and Neurotics*. London: Routledge.

Fuery, Patrick. 1995. *The Theory of Absence: Subjectivity, Signification, and Desire*. Westport, CT: Greenwood Press.

Gammage, Bill. 2002. 'Australia under Aboriginal Land Management'. Fifteenth Barry Andrews Memorial Lecture. University College, Australian Defence Force Academy, Canberra.

Gauchet, Marcel. 1997. *The Disenchantment of the World: A Political History of Religion*. Translated by Oscar Burge, with a foreword by Charles Taylor. Princeton, NJ: Princeton University Press.

Gould, Richard. 1969. 'Subsistence Behaviour among the Western Desert Aborigines of Australia'. *Oceania* 39 (4): 253–74. doi.org/10.1002/j.1834-4461. 1969.tb01026.x.

Gould, Richard A. 1982. 'To Have and Have Not: The Ecology of Sharing among Hunter-Gatherers'. In *Resource Managers: North American and Australian Hunter-Gatherers*, edited by Nancy M Williams and Eugene S Hunn, 69–98. Canberra: Australian Institute of Aboriginal Studies. doi.org/10.4324/9780429304569-4.

Graham, Mary. 1999. 'Some Thoughts about the Philosophical Underpinnings of Aboriginal Worldviews'. *Worldviews: Global Religions, Culture, and Ecology* 3: 105–18. doi.org/10.1163/156853599X00090.

Grau, Andrée. 2005. 'When the Landscape Becomes Flesh: An Investigation into Body Boundaries with Special Reference to Tiwi Dance and Western Classical Ballet'. *Body & Society* 11 (4): 141–63. doi.org/10.1177/1357034X05058024.

Gregory, Augustus C. 1884 [1981]. 'North Australia Expedition, 1865–66'. In *Journals of Australian Explorations,* Augustus C Gregory and Francis T Gregory, 99–194. Brisbane: James C Beal, Government Printer, 1884 (Facsimile edition, Hesperian Press, Perth, 1981).

Grosz, Elizabeth. 1989. *Sexual Subversions: Three French Feminists.* Sydney: Allen & Unwin.

Grosz, Elizabeth. 1994. *Volatile Bodies: Toward a Corporeal Feminism.* Bloomington, IN: Indiana University Press.

Gurevitch, Aron. 1985. *Categories of Medieval Culture.* London: Routledge and Kegan Paul.

Guss, David M. 1985. *The Language of the Birds: Tales, Texts, & Poems of Interspecies Communication.* San Francisco, CA: North Point Press.

Hallam, Sylvia. 1987. 'Changing Landscapes and Societies: 15,000 to 6000 Years Ago'. In *Australians to 1788,* edited by Dereck Mulvaney and J Peter White, 46–73. Sydney: Fairfax, Syme and Weldon Associates.

Hamilton, Annette, 1981. *Nature and Nurture: Aboriginal Child-Rearing in North-Central Arnhem Land.* Canberra: Australian Institute of Aboriginal Studies.

Hamilton, Annette. 1982. 'The Unity of Hunting and Gathering Societies: Reflections on Economic Forms and Resource Management'. In *Resource Managers: North American and Australian Hunter-Gatherers*, edited by Nancy M Williams and Eugene S Hunn, 229–48. Boulder, CO: Westview Press for the American Association for the Advancement of Science. doi.org/10.4324/9780 429304569-12.

Hamilton, Annette. 1986. 'Daughters of the Imaginary'. *Canberra Anthropology* 9 (2): 1–25. doi.org/10.1080/03149098609508533.

Handelman, Susan A. 1982. *The Slayers of Moses: The Emergence of Rabbinic Interpretation in Modern Literary Theory.* Albany, NY: State University of New York Press. doi.org/10.1353/book10773.

Harries-Jones, Peter. 1995. *A Recursive Vision: Ecological Understanding and Gregory Bateson.* Toronto: University of Toronto Press. doi.org/10.3138/9781442670440.

Head, Lesley. 1994. 'Landscapes Socialised by Fire: Post-Contact Changes in Aboriginal Fire Use in Northern Australia and Implications for Prehistory'. *Archaeology in Oceania* 29 (3): 172–81. doi.org/10.1002/arco.1994.29.3.172.

Healy, Chris. 1999. 'White Feet and Black Trails: Travelling Cultures at the Lurujarri Trail'. *Postcolonial Studies* 2 (1): 55–73. doi.org/10.1080/13688799989896.

Hoffmeyer, Jesper. 1993. *Signs of Meaning in the Universe*. Bloomington, IN: Indiana University Press.

Holmes, John H. 1990. 'Ricardo Revisited: Submarginal Land and Non-Viable Cattle Enterprises in the Northern Territory Gulf District'. *Journal of Rural Studies* 6 (1): 45–65. doi.org/10.1016/0743-0167(90)90028-7.

Hoogenraad, Robert, and George Jampijina Robertson. 1997. 'Seasonal Calendars from Central Australia'. In *Windows on Meteorology: Australian Perspective*, edited by Eric K Webb, 34–42. Melbourne: CSIRO Publishing.

Horton, David. 1982. 'The Burning Question: Aborigines, Fire and Australian Ecosystems'. *Mankind* (renamed *Australian Journal of Anthropology* in 1990) 13 (3): 237–52. doi.org/10.1111/j.1835-9310.1982.tb01234.x.

Horton, David. 2000. *The Pure State of Nature: Sacred Cows, Destructive Myths and the Environment*. Sydney: Allen & Unwin.

Horwitz, Pierre, and Brenton Knott. 1995. 'The Distribution and Spread of the Yabby (*Cherax destructor*) Complex in Australia: Speculations, Hypotheses and the Need for Research'. *Freshwater Crayfish* 10: 81–91.

Hynes, Anthony, and Athol Chase. 1982. 'Plants, Sites and Domiculture: Aboriginal Influence upon Plant Communities in Cape York Peninsula'. *Archaeology in Oceania* 17 (1): 38–50. doi.org/10.1002/j.1834-4453.1982.tb00037.x.

Ingold, Tim. 1994. *From Trust to Domination: An Alternative History of Human–Animal Relations*. London: Routledge.

Ingold, Tim, ed. 1996. *Key Debates in Anthropology*. London: Routledge.

Ingold, Tim. 2000. *The Perception of the Environment: Essays on Livelihood, Dwelling and Skill*. London: Routledge.

Irigaray, Luce. 1991. 'Questions to Emmanuel Levinas on the Divinity of Love'. In *Re-Reading Levinas*, edited by Robert Bernasconi and Simon Critchley, 109–18. Bloomington, IN: Indiana University Press. doi.org/10.5040/9781 472547354.ch-006.

Jones, Rhys. 1969. 'Fire-stick Farming'. *Australian Natural History* 16: 224–28.

Jones, Rhys. 1995. 'Mindjongork: Legacy of the Firestick'. In *Country in Flames: Proceedings of the 1994 Symposium on Biodiversity and Fire in North Australia*, edited by Deborah Bird Rose, 11–17. Biodiversity Series 3. Canberra: Biodiversity Unit, Department of the Environment, Sport and Territories; Darwin: North Australia Research Unit, ANU. Online: openresearch-repository.anu.edu.au/ bitstream/1885/282735/1/b18944231.pdf, accessed 31 January 2023.

Kaberry, PM [Phyllis Mary]. 1939. *Aboriginal Woman: Sacred and Profane.* London: G. Routledge.

Keen, Ian. 1993. 'Review Article: Tony Swain. "A Place for Strangers. Towards a History of Australian Aboriginal Being"'. *Australian Journal of Anthropology* 4 (2): 96–110.

Kimber, Richard G. 1976. 'Beginnings of Farming? Some Man–Plant–Animal Relationships in Central Australia'. *Mankind* (in 1990 renamed *The Australian Journal of Anthropology*) 10 (3): 142–51.

Kimber, Richard G. 1983. 'Black Lightning: Aborigines and Fire in Central Australia and the Western Desert'. *Archaeology in Oceania* 18 (1): 38–45. doi.org/10.1002/arco.1983.18.1.38.

Kimber, Richard G, and Mike Smith. 1987. 'An Aranda Ceremony'. In *Australians to 1788,* edited by Derek John Mulvaney and J Peter White, 221–37. Sydney: Fairfax, Syme and Weldon Associates.

Kolig, Eric. 1981. *The Silent Revolution: The Effects of Modernization on Australian Aboriginal Religion.* Philadelphia: Institute for the Study of Human Issues.

Larson, Helen K, and Keith C Martin. 1990. *Freshwater Fishes of the Northern Territory.* Darwin: Northern Territory Museum of Arts and Sciences.

Latz, PK [Peter Kenneth]. 1995. *Bushfires and Bush Tucker: Aboriginal Plant Use in Central Australia.* Illustrated by Jenny Green. Alice Springs, NT: Institute for Aboriginal Development (IAD) Press.

Lessa, William A, and Evon Voigt. 1979. *Reader in Comparative Religion: An Anthropological Approach.* New York, NY: Harper and Row Publishers.

Letts, Goff. 1969. 'Erosion – 20 Percent of VRD Already Hit – Letts'. *NT News,* 23 August, 9.

Levinas, Emmanuel. 1994. 'Liberté et Commandement [Freedom and Command]'. In *Les imprévus de l'histoire [Unforeseen History],* edited by Pierre Hayat. Montpellier: Fata Morgana.

Levinas, Emmanuel. 1996. 'Transcendence and Height'. In *Basic Philosophical Writings,* edited by Emmanuel Levinas, 11–32. Bloomington, IN: Indiana University Press.

Lévi-Strauss, Claude. 1963. *Totemism.* Boston, MA: Beacon Press.

Lewis, Darrell. 1988. 'Hawk Hunting Hides in the Victoria River District'. *Australian Aboriginal Studies* 2: 74–78.

Lewis, Darrell. 1996. *The Boab Belt: A Survey of Historic Sites in the North-Central Victoria River District.* Report prepared for the Australian National Trust (NT).

Lewis, Darrell. 2002. *Slower than the Eye Can See: Environmental Change in North Australia's Cattle Lands. A Case Study from the Victoria River District, Northern Territory.* Darwin: Tropical Savannas CRC.

Lewis, Darrell. 2007. *The Murranji Track: 'Ghost Road of the Drovers'.* Rockhampton: Central Queensland University Press.

Lewis, Darrell. 2012. *A Wild History: Life and Death on the Victoria River Frontier.* Clayton, VIC: Monash University Publishing.

Lewis, Darrell, and Deborah Bird Rose. 1988. *The Shape of the Dreaming. The Cultural Significance of Victoria River Rock Art.* Canberra: Australian Institute of Aboriginal Studies.

Lewis, Henry T. 1993. 'In Retrospect'. In *Before the Wilderness: Environmental Management by Native Californians,* edited by Thomas C Blackburn and Kat Anderson, 389–400. Menlo Park, CA: Ballena Press.

Lines, William M. 1991. *Taming the Great South Land: A History of the Conquest of Nature in Australia.* Sydney: Allen & Unwin.

Lourandos, Harry. 1997. *Continent of Hunter-Gatherers: New Perspectives in Australian Prehistory.* Melbourne: Cambridge University Press.

Lovelock, James E. 1979. *Gaia: A New Look at Life on Earth.* Oxford: Oxford University Press.

Low, William, William Dobbie and Lisa Roeger. 1988. *Resource Appraisal of Victoria River Downs Station Pastoral Lease 680.* Prepared for the Conservation Commission of the Northern Territory, Alice Springs, by WA Low Ecological Services, Alice Springs.

Malinowski, Bronisław. 1948. *Magic, Science and Religion, and other Essays.* London: Souvenir Press.

Marshall, Alan, ed. 1966. *The Great Extermination: A Guide to Anglo-Australian Cupidity, Wickedness & Waste.* London: Heinemann.

Martin, Bernard. 1970. *Great Twentieth Century Jewish Philosophers: Shestov, Rosenzweig, Buber.* New York, NY: Macmillan Company.

Mathews, Freya. 1991. *The Ecological Self.* London: Routledge.

Maze, Wilson Harold. 1945. 'Settlement in the Eastern Kimberleys, Western Australia'. *Australian Geographer* 5 (1): 1–19. doi.org/10.1080/00049184508702244.

Melville, Ian R. 1981. *A Guide to Property Planning and Improved Pastures for the Control of Erosion in the Top End of the Northern Territory*. Darwin: Land Conservation Unit, Conservation Commission of the Northern Territory.

Merchant, Carolyn. 1980. *The Death of Nature: Women, Ecology, and the Scientific Revolution*. San Francisco, CA: Harper Collins.

Merlan, Francesca. 1988a. 'Review of John P. Lea's Government and the Community in Katherine, 1937–78'. In *Aboriginal History* (special volume in honour of Diane Barwick) 12 (1/2): 219–21.

Merlan, Francesca. 1988b. 'Review of J.S. Wolfe's Pine Creek Aborigines and Town Camps'. In *Aboriginal History* (special volume in honour of Diane Barwick) 12 (1/2): 221–22.

Michaels, Eric. 1993. *Bad Aboriginal Art: Tradition, Media and Technological Horizons*. Minneapolis, MN: University of Minnesota Press.

Midgley, Mary B. 1992. *Science as Salvation: A Modern Myth and Its Meaning*. London: Routledge.

Miller, Linda I, and Michael Crothers. 1998. 'The Northern Territory Noxious Weeds Act'. Agnote (Darwin) No. 566 7. Department of Industries and Fisheries, Northern Territory, Darwin.

Morton, Stephen R. 1990. 'The Impact of European Settlement on the Vertebrate Animals of Arid Australia: A Conceptual Model'. *Proceedings of the Ecological Society of Australia* 16: 201–13. Online: www.researchgate.net/publication/2771 43330_Morton_S_R_The_impact_of_European_settlement_on_the_vertebrate _animals_of_arid_Australia_a_conceptual_model_Proceedings_of_the_ Ecological_Society_of_Australia, accessed 30 January 2023.

Muecke, Stephen. 1999. 'The Sacred in History'. *Humanities Research* 1: 27–37. doi.org/10.22459/HR.01.1999.02.

Mulvaney, Dereck. 1987. 'The End of the Beginning: 6000 Years Ago to 1788'. In *Australians to 1788,* edited by Dereck Mulvaney and J Peter White, 7–113. Sydney: Fairfax, Syme & Weldon Associates.

Newsome, Alan. 1980. 'The Eco-Mythology of the Red Kangaroo in Central Australia'. *Mankind* 12 (4): 327–33. doi.org/10.1111/j.1835-9310.1980.tb 01207.x.

Newton, Adam. 1995. *Narrative Ethics*. Cambridge, MA: Harvard University Press.

Oppenheim, Michael D. 1997. *Speaking/Writing of God: Jewish Philosophical Reflections on the Life with Others*. SUNY Series in Jewish Philosophy. Albany, NY: SUNY Press. doi.org/10.1353/book10227.

Peterson, Nicolas. 1972. 'Totemism Yesterday: Sentiment and Local Organisation among the Australian Aborigines'. *Man* (New Series) 7 (1): 12–32. doi.org/10.2307/2799853.

Peterson, Nicolas. 1976. 'The Natural and Cultural Areas of Aboriginal Australia: A Preliminary Analysis of Population Groupings with Adaptive Significance'. In *Tribes and Boundaries in Australia,* edited by Nicolas Peterson, 50–71. Social Anthropology Series No. 10. Canberra: Australian Institute of Aboriginal Studies.

Peterson, Nicolas, and Jeremy Long. 1986. *Australian Territorial Organization*. Oceania Monograph 30. Sydney: University of Sydney.

Petheram, Richard J, and Bernard Kok. 1983. *Plants of the Kimberley Region of Western Australia*. Perth, WA: University of Western Australia Press.

Plumwood, Val. 1990. 'Plato and the Bush: Philosophy and the Environment in Australia'. *Meanjin* 49 (3): 524–36.

Plumwood, Val. 1993. *Feminism and the Mastery of Nature*. London: Routledge.

Pyne, Stephen J. 1991. *Burning Bush: A Fire History of Australia*. New York, NY: Holt.

Radcliffe-Brown, AR [Alfred Reginald]. 1929. 'Notes on Totemism in Eastern Australia'. *Journal of the Royal Anthropological Institute of Great Britain and Ireland* 59 (July–December): 399–415. doi.org/10.2307/2843892.

Rawling, Jim. 1987. 'Capital, the State and Rural Land Holdings: The Example of the Pastoral Industry in the Northern Territory'. *Australian Geographer* 18 (1): 23–32. doi.org/10.1080/00049188708702923.

Rifkin, Jeremy. 1992. *Beyond Beef: The Rise and Fall of the Cattle Culture*. New York, NY: Dutton.

Robin, Libby. 2005. 'Migrants and Nomads: Seasoning Zoological Knowledge in Australia'. In *A Change in the Weather: Climate and Culture in Australia,* edited by Tim Sherratt, Tom Griffiths and Libby Robin, 42–53. Canberra: National Museum of Australia Press.

Rolls, Eric. 1981. *A Million Wild Acres*. Ringwood, VIC: Penguin Books.

Rolls, Eric. 1984. *They All Ran Wild: The Story of Pests on the Land in Australia*. Sydney: Angus & Robertson.

Rose, Deborah Bird. 1984. 'The Saga of Captain Cook: Morality in Aboriginal and European Law'. *Aboriginal Studies* 2: 24–39.

Rose, Deborah Bird. 1991. *Hidden Histories: Black Stories from Victoria River Downs, Humbert River, and Wave Hill Stations, North Australia*. Canberra: Aboriginal Studies Press.

Rose, Deborah Bird. 1992. *Dingo Makes Us Human: Life and Land in an Australian Aboriginal Culture*. Cambridge: Cambridge University Press.

Rose, Deborah Bird. 1996. *Nourishing Terrains: Australian Aboriginal Views of Landscape and Wilderness*. Canberra: Australian Heritage Commission. Online: www.ceosand.catholic.edu.au/catholicidentity/index.php/sustainability/sustainability-and-aboriginal-education/91-nourishing-terrains/file, accessed 30 January 2023.

Rose, Deborah Bird. 1997. 'Common Property Regimes in Aboriginal Australia: Totemism Revisited'. In *The Governance of Common Property in the Pacific Region*, edited by Peter Larmour, 127–43. Canberra: National Centre for Development Studies.

Rose, Deborah Bird. 1998. 'Consciousness and Responsibility in an Australian Aboriginal Religion'. *Nelen Yubu*. Dickson, ACT: Daramalan College. Online: misacor.org.au/images/Documents/NelenYubu/NY23.PDF, accessed 30 July 2023.

Rose, Deborah Bird. 2000. 'To Dance with Time: A Victoria River Aboriginal Study'. *Australian Journal of Anthropology* 11 (2): 287–96. doi.org/10.1111/j.1835-9310.2000.tb00044.x.

Rose, Deborah Bird. 2002. 'Love and Reconciliation in the Forest: A Study in Decolonisation'. Hawke Institute Working Papers Series No. 19. Adelaide: University of South Australia.

Rose, Deborah Bird with Sharon D'Amico, Nancy Daiyi, Kathy Deveraux, Margy Daiyi, Linda Ford and April Bright. 2002. *Country of the Heart: An Indigenous Australian Homeland*. Canberra: Aboriginal Studies Press.

Rosengarten, Marsha. 1996. 'Blood and the Fragility of Identity'. PhD thesis. Sydney: University of Technology.

Rosengarten, Marsha. 2000. 'Thinking Menstrual Blood'. *Australian Feminist Studies* 15 (31): 91–101. doi.org/10.1080/713611919.

Russell-Smith, Jeremy, and David MJS Bowman. 1992. 'Conservation of Monsoon Rainforest Isolates in the Northern Territory, Australia'. *Biological Conservation* 59 (1): 51–63. doi.org/10.1016/0006-3207(92)90713-w.

Schultz, Charlie, and Darrell Lewis. 1995. *Beyond the Big Run: Station Life on Australia's Last Frontier.* St Lucia: Queensland University Press.

Scott, Colin. 1996. 'Science for the West, Myth for the Rest? The Case of James Bay Cree Knowledge Construction'. In *Naked Science: Anthropological Inquiries into Boundaries, Power and Knowledge,* edited by Laura Nader, 69–86. London: Routledge.

Scott, James. 1998. *Seeing Like a State: How Certain Schemes to Improve the Human Condition Have Failed.* New Haven, CT: Yale University Press.

Shestov, Lev. 1968. *Athens and Jerusalem.* Translated by Bernard Martin. New York, NY: Simon and Schuster.

Shestov, Lev. 1970. 'Children and Stepchildren of Time: Spinoza in History'. In *A Shestov Anthology,* edited by Bernard Martin. Athens, OH: Ohio University Press.

Smith, Nicholas, Bobby Wididburu, Roy Harrington and Glenn Whightman. 1993. *Ngarinyman Ethnobotany: Aboriginal Plant Use from the Victoria River Area Northern Australia.* Northern Territory Botanical Bulletin No. 16. Darwin: Conservation Commission of the Northern Territory.

Solway, Jaqueline S, and Richard B Lee. 1990. 'Foragers, Genuine or Spurious? Situating the Kalahari San in History'. *Current Anthropology* 31 (2): 109–46. doi.org/10.1086/203816.

Spencer, Walter Baldwin, and Francis J Gillen. 1899. *The Native Tribes of Central Australia.* London: Macmillan and Co. Ltd.

Springett, BP. 1986. 'A Nation of Trees: Australian Forest Ecosystems'. In *A Natural Legacy. Ecology in Australia,* edited by Harry F Recher, Daniel Lunney and Irina Dunn, 9–103. Sydney: Pergamon Press.

Stanner, William EH. 1933. 'Ceremonial Economics of the Mulluk Mulluk and Madngella Tribes of the Daly River, North Australia. A Preliminary Paper'. *Oceania* 4 (2): 156–75. doi.org/10.1002/j.1834-4461.1933.tb00098.x.

Stanner, William EH. 1965. 'Aboriginal Territorial Organization: Estate, Range, Domain and Regime'. *Oceania* 36 (1): 1–26. doi.org/10.1002/j.1834-4461.1965.tb00275.x.

Stanner, William EH. 1979 [1962]. 'Religion, Totemism and Symbolism'. In *White Man Got No Dreaming: Essays 1938–1973,* 106–43. Canberra: Australian National University Press. Online: hdl.handle.net/1885/114726, accessed 31 January 2023.

Stebbings, Vanessa. n.d. 'Fighting the Worst Mammal Extinction Rate in the World'. *Taronga.* Online: taronga.org.au/news/2018-07-11/fighting-worst-mammal-extinction-rate-world, accessed 13 February 2023.

Stevens, Frank. 1974. *Aborigines in the Northern Territory Cattle Industry.* Canberra: Australian National University Press.

Stevenson, Paul Murray. 1985. 'Traditional Aboriginal Resource Management in the Wet-Dry Tropics: Tiwi Case Study'. *Proceedings of the Ecological Society of Australia* 13: 309–15.

Stoneham, Gary, and Joe Johnston. 1987. *The Australian Brucellosis and Tuberculosis Eradication Campaign: An Economic Evaluation of Options for Finalising the Campaign in Northern Australia.* Canberra: Australian Government Publishing Service.

Stotz, Gertrude. 1993. 'Kurdungurlu Got to Drive Toyota: Differential Colonizing Process among the Warlpiri'. PhD thesis. Deakin University. Online: hdl.handle.net/10536/DRO/DU:30023333, accessed 13 March 2023.

Strehlow, Theodor George Henry. 1970. 'Geography and the Totemic Landscape in Central Australia: A Functional Study'. In *Australian Aboriginal Anthropology,* edited by Ronald M Berndt, 92–140. Canberra: Australian Institute of Aboriginal Studies and Perth: University of Western Australia.

Sutton, Peter. 1988. *Dreamings: The Art of Aboriginal Australia.* Ringwood, VIC: Viking, Penguin Books.

Swain, Tony. 1993. *A Place for Strangers: Towards a History of Australian Aboriginal Being.* New York, NY: Cambridge University Press. doi.org/10.1017/CBO9780511552175.

Swan, Gerry. 1995. *A Photographic Guide to Snakes and Other Reptiles of Australia.* Frenchs Forrest, NSW: New Holland Publishers.

Tindale, Norman B. 1974. *Aboriginal Tribes of Australia: Their Terrain, Environmental Controls, Distribution, Limits, and Proper Names.* 2 parts. Canberra: Australian National University Press. Online: hdl.handle.net/1885/114913, accessed 31 January 2023.

Toop, Carla R. 1958. 'Kimberley Horse Disease ("Walkabout Disease")'. *Journal of the Department of Agriculture, Western Australia,* Series 3, 7 (4) (July–August): 398–403. Online: researchlibrary.agric.wa.gov.au/journal_agriculture3/vol7/iss4/4, accessed 30 January 2023.

Tuan, Yi-Fu. 1977. *Space and Place: The Perspective of Experience.* London: Edward Arnold.

Turner, David H. 1996. *Return to Eden: A Journey through the Aboriginal Promised Landscape of Amagalyuagba*. New York, NY: P. Lang.

Tyler, Stephen A. 1984. 'The Vision Quest in the West, or What the Mind's Eye Sees'. *Journal of Anthropological Research* 40 (1): 23–40. doi.org/10.1086/jar.40.1.3629688.

United Nations. 1992. *United Nations Conference on Environment and Development (UNCED)*. Rio de Janeiro, Brazil, 3–14 June. Online: www.un.org/en/conferences/environment/rio1992, accessed 21 February 2023.

Wagner, Roy. 1975. *The Invention of Culture*. Englewood Cliffs, NJ: Prentice-Hall.

Walden, David, Rik van Dam, Max Finlayson, Michael Storrs, John Lowry and Darren Kriticos. 2004. *A Risk Assessment of the Tropical Wetland Weed* Mimosa pigra *in Northern Australia*. Supervising Scientist Report 177. Darwin, NT: Supervising Scientist. Online: www.dcceew.gov.au/sites/default/files/documents/ssr177-print.pdf, accessed 30 January 2023.

Walsh, Fiona J, B Cross and members of Wangkatjungka Aboriginal Corporation, Kupartiya Incorporated, Ngumpan Aboriginal Corporation, Dodnun community. 2003. *Bushfires and Burning – Aspects of Aboriginal Knowledge and Practice in Areas of the Kimberley*. Report to Kimberley Regional Fire Management Project, Broome, WA.

Warner, W Lloyd. 1958 (1937). *A Black Civilization: A Social Study of an Australian Tribe*. New York, NY: Harper and Brothers.

Watson, Pamela. 1983. *This Precious Foliage: A Study of the Aboriginal Psycho-Active Drug Pituri*. Oceania Monograph 26. Sydney: University of Sydney.

Whitford, Margaret. 1991. *Luce Irigaray: Philosophy in the Feminine*. London: Routledge. doi.org/10.4324/9781315824741.

Wightman, Glenn M. 1994. *Gurindji Ethnobotany: Aboriginal Plant Use from Daguragu, Northern Australia*. Darwin: Conservation Commission of the Northern Territory.

Wild, Stephen. 1977–78. 'Men as Women: Female Dance Symbolism in Warlbiri Men's Rituals'. *Dance Research Journal* 10 (1): 14–22. doi.org/10.2307/1478492.

Wild, Stephen A. 1984. 'Warlbiri Music and Culture: Meaning in a Central Australian Song Series'. In *Problems and Solutions: Occasional Essays in Musicology, Presented to Alice M. Moyle,* edited by Jamie C Kassler and Jill Stubington, 186–203. Sydney: Hale & Iremonger.

Williams, Nancy, and Eugene S Hunn, eds. 1982. *Resource Managers: North American and Australian Hunter-Gatherers*. Canberra: Australian Institute of Aboriginal Studies.

Williams, Nancy, and Graham Baines, eds. 1993. *Traditional Ecological Knowledge: Wisdom for Sustainable Development*. Centre for Resource and Environmental Studies. Canberra: The Australian National University.

Willshire, William H. 1896. *The Land of Dawning, Being Facts Gleaned from Cannibals in Australian Stone Age*. Adelaide: W.K. Thomas.

Wilson, James S. 1885. 'Notes on the Physical Geography of North-West Australia'. *Journal of the Royal Geographical Society of London* 28: 137–53. Online: nla. gov.au/nla.obj-61947043/view?partId=nla.obj-61949285#page/n0/mode/1up, accessed 30 January 2023.

Woinarski, John CZ, Andrew A Burbidge and Peter L Harrison. 2012. 'Ongoing Unraveling of a Continental Fauna: Decline and Extinction of Australian Mammals since European Settlement'. *PNAS* 112 (15): 4531–40. doi.org/ 10.1073/pnas.1417301112.

Wolfe, Patrick. 1999. *Settler Colonialism and the Transformation of Anthropology: The Politics and Poetics of an Ethnographic Event*. London: Cassell.

Wolfe, Patrick. 2000. 'White Man's Flour: The Politics and Poetics of an Anthropological Discovery'. In *Colonial Subjects: Essays on the Practical History of Anthropology*, edited by Peter Pells and Oscar Salemink, 205–31. Ann Arbor, MI: University of Michigan Press.

Worsley, Peter. 1967. 'Groote Eylandt Totemism and Le Totémisme aujourd'hui'. In *The Structural Study of Myth and Totemism,* edited by Edmund Leach, 141–60. London: Tavistock.

Yunupingu, Joe. 1995. 'Fire in Arnhem Land'. In *Country in Flames: Proceedings of the 1994 Symposium on Biodiversity and Fire in North Australia,* edited by Deborah Bird Rose, 65–66. Biodiversity Series 3. Canberra: Biodiversity Unit, Department of the Environment, Sport and Territories; Darwin: The North Australia Research Unit, ANU. Online: openresearch-repository.anu.edu.au/bitstream/1885/282 735/1/b18944231.pdf, accessed 31 January 2023.

Zimmerman, Frank. 1996. *Why Haldane Went to India: Modern Genetics in Quest of Tradition*. Oxford: Oxford University Press. doi.org/10.1093/acprof:oso/ 9780198288848.003.0008.

Legislation

Aboriginal Land Rights (Northern Territory) Act 1976. Online: www.foundingdocs. gov.au/item-did-57.html, accessed 3 February 2023.

Noxious Weeds Act 1963 (Northern Territory of Australia; since repealed). Online: www.ecolex.org/details/legislation/noxious-weeds-act-lex-faoc018547/, accessed 7 March 2023.

Soil Conservation and Land Utilisation Act 1969 (Northern Territory of Australia). Online: legislation.nt.gov.au/en/Legislation/SOIL-CONSERVATION-AND-LAND-UTILISATION-ACT-1969, accessed 13 September 2023.

Index

Note: page numbers in *italics* indicate information to be found in a table or figure. Page numbers in **bold** indicate information to be found in a block quote from one of Rose's teachers.

www.ingramcontent.com/pod-product-compliance
Lightning Source LLC
Chambersburg PA
CBHW051950270326
41929CB00015B/2602